Knowledge-Based Driver Assistance Systems

Michael Huelsen

Knowledge-Based Driver Assistance Systems

Traffic Situation Description
and Situation Feature Relevance

Michael Huelsen
Karlsruhe, Germany

Also PhD Thesis Karlsruhe Institute of Technology (KIT), Germany, 2013

ISBN 978-3-658-05749-7   ISBN 978-3-658-05750-3 (eBook)
DOI 10.1007/978-3-658-05750-3

The Deutsche Nationalbibliothek lists this publication in the Deutsche Nationalbibliografie; detailed bibliographic data are available in the Internet at http://dnb.d-nb.de.

Library of Congress Control Number: 2014938109

Springer Vieweg
© Springer Fachmedien Wiesbaden 2014

Printed on acid-free paper

Springer Vieweg is a brand of Springer DE.
Springer DE is part of Springer Science+Business Media.
www.springer-vieweg.de

Acknowledgment

This thesis was created under the roof of a fruitful cooperation with KIT Karlsruhe Institute of Technology, Faculty of Informatics and Robert Bosch GmbH, Leonberg.

In the first place I would like to thank Prof. Dr. Marius Zöllner for his support and supervision of this thesis and his whole team for the consecutive and sustainable availability for stimulating research discussions and guidance throughout the process of creation of this thesis. Furthermore, many thanks go to Prof. Dr. Rudi Studer for acting as second examiner. Particular thanks also go to Dr. Marcus Strand, Thomas Schamm, Dennis Nienhüser, Tobias Bär, Thomas Gumpp, Sebastian Brechtel and many more.

I equally appreciate the cooperation within the Robert Bosch GmbH, particularly the department of radar-based driver assistance systems within the division Chassis Control Systems. This cooperation not only enabled deep insight into real world systems and data and the test and development of new methods based on such, but also to extend scientific and pragmatic exchange to the industry and frontier of product development. Many thanks therefore particularly go to Ulrich Sailer, Lars Berding, Dr. Christian Weiß, Dr. Ulrike Ahlrichs, Stephanie Arramon, Esther Wenzel, Dr. Joachim Börger, Dr. Jürgen Häring, Dr. Björn Fassbender, Christian Niebuhr, Jacek Kaminski, Michael Egert, Heiko Kirn, Raphael van Uffelen, Andres Grimm, Stephane Preau and many more.

Jakob Spinda, Johannes Hesser and Nino Häberlein significantly supported this work with their theses and internship for implementation and testing of methods and algorithms developed with this thesis.

My utmost thanks go to Venelin Staynov and Robert Scholz for their everlasting and motivating friendship, through even harsh times.

Last but certainly above all, I appreciate my family to always have been there with and for me, for enabling my comprehensive studies, for supporting my work and my aims and to offer their open hand, when needed.

Michael Huelsen

Abstract

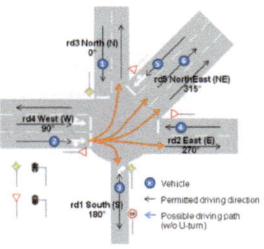 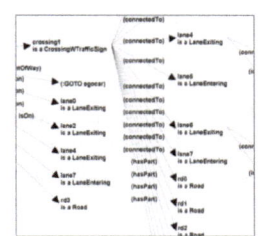 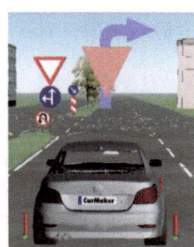

The comprehension of a traffic situation plays a major role in driving a vehicle. It transforms perceived raw information into interpretable information. This forms a basis for future projection, decision making and action performing, such as navigating, maneuvering and driving control, within the driving control loop.

The aim of this thesis is to provide a generic traffic situation description capable of supplying various ADAS with relevant information about the current driving and traffic situation of the ego vehicle and its environment. With this information ADAS should be able to perform reasonable functions and actions and approach visionary goals such as injury and accident free driving, substantial assistance in arbitrary situations up to even autonomous driving.

Knowledge-based Traffic Situation Description

Most complex traffic situations seem to be those at intersections. Their understanding is influenced by a variety of object and relation types such as intersecting roads with lanes and markings, allowed and forbidden paths, vehicles coming from different directions and different kinds of road signs (see figure on the top left). Ontologies are well suited for modeling these multi-object traffic situations and for performing logic reasoning to check consistency of its knowledge and to reason about object types, relations and to e.g. apply traffic rules. *Description logic* (DL) is a language for building such ontologies

Key Contributions

Three key contributions are provided by this thesis in the context of advanced driver assistance systems, more precisely to efficiently describe the current traffic situation a vehicle is part of:

1. The empirical proof that *mutual information based feature selection* serves to effectively evaluate the relevance of or select *traffic situation features*, especially those provided by ADAS sensors and ADAS sensor data fusion.

2. Theoretical discussion and empirical proof that *description logic based ontologies* are suitable to describe *complex traffic situations* such as those at intersections, including the evaluation of traffic rule compliance.

3. Theoretical discussion and empirical proof that *mutual information based feature selection* may be utilized on *semantic features* contained in *description logic based ontologies* for a more comprehensive and lean traffic situation description.

Mutual Information Based Traffic Situation Feature Selection

As the first main contribution, tremendous benefit in performance and time consumption of MI feature selection could be shown on vast, real world measurement data compared to feature selection by experts working in serial function development. Additionally, a methodical improvement to existing methods including formalization and error calculations is given and empirically shown. Another contribution is an approach and its interpretation for the usage of mutual information to determine the relevance of the history of situation features.

Description Logic Based Traffic Situation Description for ADAS

This thesis is widely concerned with the investigation and validation of applicability and capability of ontology based traffic situation description. Knowledge engineering for complex traffic situations such as those at complex intersections is proven to be feasible. Furthermore, the developed description logic based ontology allows logic reasoning of traffic rules.

Additionally, this thesis provides a proof-by-implementation for logic-based situation description for real-time execution of driver assistance functions. An asynchronous real-time framework is used, especially designed for the herewith proposed ontological situation description. It is usable for arbitrary DAS functions; shown on exemplary DAS functions.

An open issue with ongoing research is coping with uncertainty in combination with knowledge-based approaches. Early approaches are too slow for implementation, especially with the number of objects and relations required for the proposed, generic situation description. This thesis introduces and discusses a concept of handling and incorporating different types of sensory uncertainty with the deterministic ontological situation description to facilitate some sensory uncertainty handling.

Mutual Information Based Selection of Ontology Features

Finally, this thesis defines semantic features derived from ontological content to perform mutual information based feature selection on ontology elements and to determine their relevance with respect to some given target application. Execution of this approach on some experimental data shows its applicability in principle.

Kurzfassung

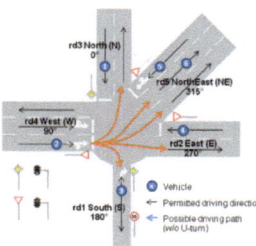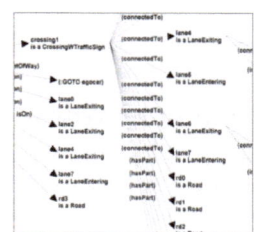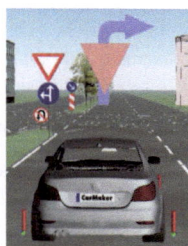

Das Verständnis einer Verkehrssituation spielt eine signifikante Rolle während der Fahrzeugführung. Es wandelt erhaltene Rohinformation in interpretierbare Informationen. Dies bildet die Grundlage für Projektion, Entscheiden und Handeln, wie Navigation, Manövrieren und Antriebssteuerung, innerhalb des fahrerischen Regelkreises.

Das Ziel dieser Arbeit ist eine generische Verkehrssituationsbeschreibung, die in der Lage ist, verschiedene Fahrerassistenzsysteme (FAS) mit relevanten Informationen über die aktuelle Fahr- und Verkehrssituation des eigenen Fahrzeugs und seiner Umgebung zu versorgen. Diese Informationen sollen FAS befähigen, sinnvolle Funktionen und Aktionen auszuführen und visionären Zielen wie unfallfreies Fahren, umfassende Unterstützung in beliebigen Situationen bis hin zum autonomen Fahren näher zu kommen.

Wissensbasierte Verkehrssituationsbeschreibung

Die komplexesten Verkehrssituationen scheinen diejenigen an Kreuzungen. Ihr Verständnis wird durch eine Vielzahl an Objekt- und Relationsarten beeinflusst, u. a. kreuzende Straßen mit Spuren und Markierungen, zugelassene und verbotene Fahrwege, aus verschiedenen Richtungen kommende Fahrzeuge und viele Arten von Verkehrsschildern (s. Abb. o. l.). Ontologien sind gut geeignet, diese Multi-Objekt-Situationen zu modellieren und durch logisches Schließen Konsistenz des enthaltenen Wissens zu prüfen und Objekttypen sowie Relationen zu ermitteln und z. B. Verkehrsregeln anzuwenden. *Description logic* (DL) ist eine Sprache, mit der solche Ontologien gebildet werden können.

Wesentliche Beiträge dieser Arbeit

Drei wesentliche Beiträge liefert diese Arbeit im Rahmen von Fahrerassistenzsystemen, präziser gesagt, um effizient die aktuelle Verkehrssituation zu beschreiben, in der sich ein Fahrzeug befindet:

1. Der empirische Nachweis, dass *Mutual-Information-basierte Merkmalsauswahl* erfolgreich dazu verwendet werden kann, die Bestimmung der Relevanz oder die Auswahl von Verkehrssituationsmerkmalen zu unterstützen, insbesondere solche von FAS-Sensoren oder FAS-Sensordatenfusion.

2. Theoretische Diskussion und empirischer Nachweis, dass *DL-basierte Ontologien* geeignet sind, *komplexe Verkehrssituationen* zu beschreiben, z. B. an Kreuzungen, einschließlich Auswertung von Verkehrsregeln und deren Einhaltung.

3. Theoretische Diskussion und empirischer Nachweis, dass *Mutual-Information-basierte Merkmalsauswahl* auf *semantische Merkmale* aus *DL-basierten Ontologien* für eine verbesserte Situationsbeschreibung angewendet werden kann.

Mutual-Information-basierte Auswahl von Verkehrssituationsmerkmalen

Als erster wesentlicher Beitrag konnten enorme Vorteile in Ergebnis und Geschwindigkeit von MI-Merkmalsauswahl gegenüber Auswahl durch Experten aus der automobilen Serienfunktionsentwicklung auf große Mengen realer Messdaten gezeigt werden. Darüber hinaus wird eine methodische Verbesserung bestehender Methoden einschließlich Formalisierung und Fehlerberechnungen gegeben und empirisch gezeigt. Ein zusätzlicher Beitrag ist ein Ansatz und seine Interpretation für die Benutzung von Mutual Information, um die Relevanz der Historie von Situationsmerkmalen zu bestimmen.

Description-Logic-basierte Verkehrssituationsbeschreibung für FAS

Diese Arbeit befasst sich ausführlich mit der Untersuchung und Validierung der Anwendbarkeit und Leistungsfähigkeit von ontologiebasierter Verkehrssituationsbeschreibung. Es wird gezeigt, dass Wissensverarbeitung für komplexe Verkehrssituationen wie z. B. an komplexen Kreuzungen realisierbar ist. Darüber hinaus ermöglicht die entwickelte DL-basierte Ontologie logisches Schlussfolgern von Verkehrsregeln.

Weitergehend bietet diese Arbeit einen praktischen Machbarkeitsnachweis für logikbasierte Situationsbeschreibung für die Echtzeitanwendung von FAS. Es wird ein asynchrones Echtzeit-Framework verwendet, das speziell für die hier vorgestellte ontologische Situationsbeschreibung entwickelt wurde. Es lässt sich für beliebige FAS-Funktionen einsetzen, wie an beispielhaften FAS-Funktionen veranschaulicht wird.

Eine offene Frage der laufenden Forschung ist die Bewältigung von Unsicherheiten in Verbindung mit wissensbasierten Ansätzen. Frühe Ansätze sind zu langsam für die Umsetzung, vor allem für die Anzahl von Objekten und Relationen, die für die vorgestellte generische Situationsbeschreibung notwendig sind. Diese Arbeit diskutiert ein Konzept zur Einbindung von verschiedenen Arten von Sensorunsicherheiten in die deterministische ontologische Situationsbeschreibung, um deren Behandlung teilweise zu ermöglichen.

Mutual-Information-basierte Auswahl von Ontologiemerkmalen

Abschließend definiert diese Arbeit semantische Merkmale, abgeleitet aus ontologischem Wissen, um eine Mutual-Information-basierte Auswahl von Ontologieelementen durchzuführen und dabei deren Relevanz bezüglich einer Zielanwendung zu bestimmen. Die Ausführung auf experimentelle Daten zeigt deren grundsätzliche Anwendbarkeit.

Table of Contents

List of Abbreviations

ABox.............. Assertional Box
ABS .... Anti Blocking System
ACC............... Adaptive Cruise Control
ADAS .... Advanced Driver Assistance System
AI.................... Artificial Intelligence
ASIL .... Automotive Safety Integrity Level
BN.................. Bayes Network, Belief Network
CAN............... Controller Area Network
CG.................. Conceptual Graph
CNF .... Conjunctive Normal Form
CPU .... Central / Computational Processing Unit
CWA............... Closed World Assumption
DARPA........... Defense Advanced Research Projects Agency
DAS .... Driver Assistance System
DL.................. Description Logic
ECU .... Electronic Control Unit
ESP .... Electronic Stability Control
FL .... Fuzzy Logic
FOL............... First Order Logic
FOPL .... First Order Probabilistic Languages
GPS............... Global Positioning System
GPU .... Graphical Processing Unit
HMI .... Human Machine Interface
HMM .... Hidden Markov Model
I2I .... Infrastructure to Infrastructure
ICA .... Independent Component Analysis
IDAF.............. Interface and Driver Assistance Module
JDL DFG........ Joint Directors of Laboratories Data Fusion Group
JMI................. Joint Mutual Information (ranking)
LCWA............ Local Closed World Assumption
MaxRel .... Maximum Relevance ranking
MCMC........... Markov Chain Monte Carlo
MCRMR .... Minimum Class Redundancy Maximum Relevance ranking
MI .... Mutual Information
MIFS.............. Mutual Information Feature Selection
MLN .... Markov Logic Network
MRF............... Markov Random Field
MRMR........... Minimum Redundancy Maximum Relevance ranking
NHTSA........... National Highway Traffic Safety Administration
NN .... Neural Network

norm................ normalized (in formula)
OOBN............. Object Oriented Bayesian Network
OPRM............. Object Oriented Probabilistic Relational Model
OWA............... Open World Assumption
OWL .... Web Ontology Language
PCA .... Principal Component Analysis
POMDP........... Partially Observable Markov Decision Process
RCC .... Region Connection Calculus
RDF .... Resource Definition Framework
rel.................... relative (in formula)
RF .... Random Forests classifier
RNDF............. Road Network Definition File
RSAP .... Road Safety Action Programme
SA, SAW .... Situation Awareness
SDF................ Sensor Data Fusion
SGT................ Situation Graph Tree
SM .... State Machine
SVM................ Support Vector Machine
TBox............... Terminological Box
TCP................ Transmission Control Protocol
TISDO............ Traffic Intersection Situation Description Ontology
TTC................ Time To Collision
TTI................. Time To Intersection
UV .... Unmanned Vehicle
V2V .... Vehicle to Vehicle
V2I................. Vehicle to Infrastructure (and vice versa)

List of Symbols

Symbol	Unit	Description
H	[bit]	Entropy
p, P	[1]	Probability of distribution, of single event
I	[bit]	Mutual information
X, Y, Z	[*any*]	Arbitrary random variables
F, C	[*any*]	Random variable of feature / class
S, F, C	[{ }]	Set of random variables
Φ	[1]	Decision function
n	[1]	Quantity
t, T, θ, Θ	[*s*]	Time
f, c	[*any*]	Signal of feature / class
C, R, R_a	[ ]	Ontological concept, role, attribute role
A	[*any*]	Ontological attribute
c	[ ]	Ontological object
r	[ ]	Ontological relation
x, e	[ ]	Ontological object or relation
$\mathcal{A}$	[{ }]	Ontological ABox
n	[{1}]	Quantity vector
P	[{1}]	Probability matrix
I	[{ }]	Attribute interval
G	[{ }]	Group of ontological elements (objects, relations)

For the notation of ontologies kindly refer to **Table 1** and **Table 2** on page 37.

Chapter 1

Introduction

1.1 Motivation

The declared objective of the European Commission was to halve the number of traffic fatalities in the period from 2001 to 2010 [European Commission, 2001]. With a decrease of 44% this goal could not completely be achieved (see **Fig. 1**). Consequently, the goal was renewed and adjusted for the period from 2011 to 2020 in [European Commission, 2010] and new strategic objects based on the current "3rd road safety action programme" (RSAP) are derived to propose new actions for the subsequent RSAP. The aim for 2020 is to "halve the overall number of road deaths in the European Union by 2020 starting from 2010".

To pursue the reduction of the number of accidents, automotive manufacturers and suppliers are conducting intensive research on *Driver Assistance Systems* (DAS). These systems support the driver in critical situations or intervene in the driving process to avoid accidents or to reduce their severity. In addition, comfort assistance functions, which make driving more convenient, are an important buying criterion to customers.

This research on ADAS is supported by [European Commission, 2010] with the main objectives n°4 "Safe vehicles" and n°5 "Promote the use of *modern technology* to increase road safety". On objective n°4, actions of the [European Commission, 2010] especially include to

> "make proposals to encourage progress on the *active* and passive *safety* of vehicles [...]" and to "further assess the impact and benefits of *co-operative systems* to identify most beneficial applications and recommend the relevant measures for their synchronized deployment."

Actions on the objective n°5 include to

> "evaluate the feasibility of retrofitting commercial vehicles and private cars with *Advanced Driver Assistance Systems*" and to "notably propose technical specifications necessary to *exchange data and information between vehicles* (V2V[1]), between vehicles *and infrastructure* (V2I[2]) and between infrastructures (I2I[3]). The possibility of extending the implementation of Advanced Driver Assistance Systems (ADAS) such as Lane Departure Warning, Anti Collision Warning or Pedestrian Recognition systems by retrofitting them to existing commercial and/or private vehicles should also be further assessed".

---

[1] Vehicle to Vehicle
[2] Vehicle to Infrastructure
[3] Infrastructure to Infrastructure

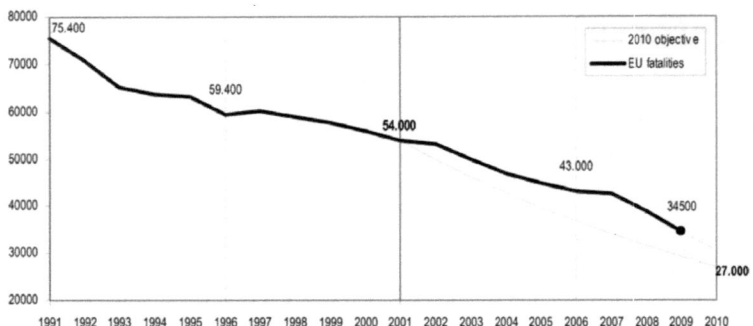

Fig. 1: Development of road traffic fatalities until 2009 (2010 estimation).
[European Commission, 2010]

The National Highway Traffic Safety Administration (NHTSA) in the U.S. pursues similar goals. A concrete plan for the years 2014 to 2020 is still under development. However, priority research fields and other significant projects with large benefits for traffic safety are, amongst others, named with forward collision avoidance and mitigation, vehicle communications, (driver) distraction, lane departure prevention, blind spot detection and pedestrian detection [NHTSA (National Highway Traffic Safety Administration), 2011].

Related projects exist, funded by the European Union, and can be found at the European Commission under the area "Road Safety Projects" within the "Transport" policy and under the "eSafety" activity within the "Information Society" department. Current examples are ADVISORS (Action for advanced Driver assistance and Vehicle control systems Implementation, Standardisation, Optimum use of the Road network and Safety), APROSYS (Advanced Protection Systems) and PReVENT (Preventive and Active Safety Applications Integrated Project) [European Commission – Information Society Technologies, 2008], the latter including INTERSAFE, WILLWARN, SAFELANE, APALACI, COMPOSE, ProFusion and more. As a successor of INTERSAFE, the INTERSAFE-2 project [INTERSAFE-2, 2011] was launched.

Advanced driver assistance systems (ADAS) being in research focus help the driver with information or by performing supportive or autonomous actions. This demands an extensive *comprehension* of the vehicle environment and *complex situations*, potentially even unknown situations. Therefore, for further *situation interpretation*, a generic method for *situation description* is needed.

Intersection assistance belongs to the field of ADAS yet to be developed, especially for in-vehicle application. Various approaches to model intersection infrastructure, geometry and dynamics exist, e.g. knowledge-based scene description, graph based methods with map data up to holistic intersection handling with state machines and trajectory planning. However, to the highest extent its complexity of situation description, real-time issues and the lack of sufficient sensory information (e.g. detailed map data, sensor accuracy and field of view) impede the start of a widely spread serial implementation in ongoing automotive development. Since intersection accidents, depending on the region, accumulate 30 to 60% of injury and up to one 3rd of fatal accidents, interest in this kind of assistance remains high [INTERSAFE-2, 2011].

1.2 Objective and Contribution of this Thesis

The comprehension of a traffic situation plays a major role in driving a vehicle. It transforms perceived raw information into interpretable information. Within the driving control loop, this forms a basis for future projection, decision making and action performing, such as navigating, maneuvering and driving control. Perception, situation comprehension and projection as well as decision making and action performing can be mapped on dynamic decision making as delineated in the model of situation awareness by [Endsley, 2000]. This applies for human decision making investigated by Endsley as well as for substantial advanced driver assistance systems (ADAS).

The aim of this thesis is to provide a generic traffic situation description capable of supplying various ADAS with relevant information about the current driving and traffic situation of the ego vehicle and its environment. With this information ADAS should be able to perform reasonable functions and actions and approach visionary goals such as injury and accident free driving, substantial assistance in arbitrary situations up to even autonomous driving.

This requires assessing more complex situations compared to state of the art assistance systems, potentially including even unknown situations demanding for extensive comprehension of the vehicle environment and the current situation.

Specific goals for the situation description are *modularity* and *hierarchy, expandability, exchangeability* and *scalability* for different variants and combinations of ADAS in vehicle applications as well as *determinism* and *correctness.*

Situation Feature Selection

As part of describing a situation, it is of interest, what information is actually relevant to target applications. For ADAS, a number of partially redundant sensors is used to generate a variety of measurement signals. These signals – or features – are then further processed, adding new features. With this high level data complexity and manifoldness, an interpretation and determination of the most relevant features becomes unwieldy. However, selecting relevant information is already asked for by Endsley concerning the perception and comprehension layers in situation awareness.

Mutual Information (MI) is a well-established tool for feature selection and feature comparison, for example, to support image classification methods or computer aided diagnosis. As an advantage, it does not preliminarily need a classifier or other machine learning algorithms for relevance calculation.

Knowledge-based Traffic Situation Description

Most complex traffic situations seem to be those at intersections. Their understanding is influenced by a variety of object and relation types such as intersecting roads with lanes and markings, allowed and forbidden paths, vehicles coming from different directions and different kinds of road signs. Their constellation directly influences traffic rules which apply and, accordingly, the assessment of allowed actions, expected behavior and impact of traffic participants among each other.

Ontologies are a foundation for knowledge representation and provide a formalism to structure objects, their relations and attributes and for performing logic reasoning with them. Therefore ontologies are well suited for modeling the kind of described multi-object traffic situations and for performing logic reasoning to check satisfiability of the situation ontology, to check consistency of input data and to reason about object types, relations and to e.g. apply traffic rules.

Description logic (DL) is a language for building ontologies and in most cases, depending on the dialect applied, it allows for decidable, complete and terminating algorithms.

Key Research Contributions within this Thesis

Three key contributions are provided by this thesis in the context of advanced driver assistance systems, more precisely to efficiently describe the current traffic situation a vehicle is part of:

1. The empirical proof that *mutual information based feature selection* serves to effectively evaluate the relevance of or select *traffic situation features*, especially those provided by ADAS sensors and ADAS sensor data fusion.

2. Theoretical discussion and empirical proof that *description logic based ontologies* are suitable to describe *complex traffic situations* such as those at intersections, including the evaluation of traffic rule compliance.

3. Theoretical discussion and empirical proof that *mutual information based feature selection* may be utilized on *semantic features* contained in *description logic based ontologies* for a more comprehensive and lean traffic situation description.

Previous work concerning contribution 1 barely exists with respect to vast real world measurement data and driver assistance. Previous work on contribution 2 does not exist when traffic rules are to be included in ontology reasoning or when real-time capability is to be shown. Previous work on contribution 3 does not exist. Existing work, the context and aspects concerning the first two contributions will be discussed throughout Chapter 2.

The following three subsections further explain the three above stated contributions of this thesis to provide a brief overview. Contribution 1 is discussed in detail in Chapter 4, contribution 2 in Chapter 5 and contribution 3 in Chapter 6.

Contribution 1: Mutual Information Based Traffic Situation Feature Selection

Initially, it is discussed how to use and interpret mutual information feature selection and what applications in the context of advanced driver assistance function development it may be used for. A methodical improvement to existing methods including formalization and error calculations is given and empirically shown with this thesis.

As the first main contribution, tremendous benefit in performance and time consumption of MI feature selection could be shown on vast, real world measurement data compared to feature selection by experts working in serial function development (see also publication [Hülsen et al., 2010]).

As the second main contribution, an approach and its interpretation for the usage of mutual information to determine the relevance of the history of situation features is developed. This approach considers past feature values laying back a certain amount of time.

Contribution 2: Description Logic Based Traffic Situation Description for ADAS

This thesis is widely concerned with the investigation and validation of applicability and capability of ontology based traffic situation description. Knowledge engineering for complex traffic situations such as those at complex intersections with an arbitrary number of roads, lanes, driving directions, allowed driving paths, traffic signs and / or lights was performed and is proved to be feasible. Little work is yet existent with this complexity of road structure including traffic infrastructure and legislation elements. No previous work exists that uses an ontology to model this kind and complexity of traffic situations. The contributed traffic situation description ontology fulfils the required aspects modularity and hierarchy, expandability, exchangeability, scalability, determinism and correctness.

Furthermore, the developed description logic based ontology allows logic reasoning of traffic rules, as has not been achieved in any previous work. Previous work struggled with a lack of reasoning capability of utilized reasoners or lack of efficient, target oriented knowledge engineering. Both the developed ontology for complex traffic situations and the inclusion of logic reasoning on traffic rules have been published in [Hülsen et al., 2011b].

Additionally, this thesis provides a proof-by-implementation for logic-based situation description for real-time execution of driver assistance functions. An asynchronous real-time framework is used, especially designed for the herewith proposed ontological situation description. It is usable for arbitrary DAS functions; shown on exemplary DAS functions (see also publication [Hülsen et al., 2011a] and associated master theses [Hesser, 2011, Spinda, 2011]). Previous work did not show generic applicability for several different DAS functions, neither in real-time, nor including reasoning about traffic rules within an ontology.

An open issue with ongoing research is coping with uncertainty in combination with knowledge-based approaches. Early approaches are too slow for implementation, especially with the number of objects and relations required for the proposed, generic situation description. This thesis introduces and discusses a concept of handling and incorporating different types of sensory uncertainty with the deterministic ontological situation description provided in this thesis to facilitate some sensory uncertainty handling.

Contribution 3: Mutual Information Based Selection of Ontology Features

Finally, this thesis defines semantic features derived from ontological content to perform mutual information based feature selection on ontology elements and to determine their relevance with respect to some given target application. Execution of this approach on some experimental data shows its applicability in principle.

1.3 Outline of this Thesis

The next **Chapter 2** briefly introduces the domain of advanced driver assistance systems (ADAS) and their components. It points out some requirements. An integral part of ADAS is situation awareness including a generic situation description. Aspects of and approaches for situation description are then discussed. Furthermore, the usefulness of determining the relevance of traffic situation features and of a knowledge-based method for traffic situation description is explained. The chapter is closed with a delimitation of the area of research relevant to this thesis with respect to the field of intelligent vehicle systems.

Chapter 3 introduces some theoretical foundations relevant to the further line of argument. This concerns mutual information feature selection for relevance determination of traffic situation features. Alongside, the Random Forests classifier is introduced as a machine learning algorithm to evaluate feature selection performance. Theoretical foundations are furthermore provided for knowledge representation, especially ontologies and description logics in combination with logic reasoning and some probabilistic knowledge-base approaches. This forms the basis for knowledge-based situation description.

Determining and interpreting the relevance of traffic situations features for target applications within the driver assistance domain is enfolded in **Chapter 4**. Besides general interpretation issues and the application on data from vehicle endurance runs, methodical improvements, relevance calculation of more than one feature at a time, as well as investigation of historical relevance of situation features are discussed in detail.

Chapter 5 deals with the approach of knowledge-based traffic situation description. The first part introduces knowledge-base engineering to model traffic situations and points out how it can be used to realize assistance functions. The second part then introduces an asynchronous real-time framework that is meant to not only simulate and test the intersection model but also to prepare in-vehicle applicability. Subsequently, the third part discusses possible assistance functions and provides simulation results for exemplary DAS functions. These results are obtained with simulations on the introduced framework utilizing the developed ontology. Finally, the fourth part addresses an approach to handle uncertainty by maintaining the determinism of the used, formal ontology.

With the ontology being part of the situation description and, in a general sense, containing complex features, **Chapter 6** proposes a method to combine mutual information based feature selection with knowledge-bases. Hence, it provides a way to determine the relevance of elements of the developed traffic situation description ontology.

Chapter 7 concludes this thesis and gives an outlook on fields of interest for further research & development and issues of a series implementation within vehicles equipped with advanced driver assistance systems.

Chapter 2

The Research Domain of this Thesis and its State of the Art

2.1 Advanced Driver Assistance Systems

Driver Assistance Systems (DAS) are additional electronic systems in vehicles to support the driver in specific driving situations. They may aim at increasing safety, higher comfort, less fuel (or resource) consumption or informing the driver about the current traffic condition, the traffic situation or the route. These systems may be informative, semi-autonomous or autonomous. They may intervene in vehicle propulsion, actuation or vehicle control or may simply provide useful additional information.

Components and architecture of an intelligent automobile and hence, the combination of a variety of DAS are described in [Stiller et al., 2007, Robert Bosch GmbH, 2011b, Robert Bosch GmbH, 2011a]. DAS are based on the control loop of mechatronic systems as defined in [DIN IEC 60050-351:2006, 2006] (formerly DIN 19226). Basically, DAS consist of multiple sensors for perception, one or several ECUs for information processing and up to several actuators for DAS function execution. This will be briefly introduced in the next subsections 2.1.1 to 2.1.3.

2.1.1 Advanced Driver Assistance Functions

DAS with various specific functions are already commonly used within vehicles. Such functions are, for example, speed control, parking assist, rain-light-sensor, anti blocking system (ABS) and electronic stability control (ESP). These systems operate with a very specific focus of control, in very specific situations or under specific conditions.

ADAS research has recently focused on providing extended functionality in a wider range of situations and with less supervision by the driver. Such functions include adaptive cruise control (ACC), (semi-) autonomous parking, pre-crash, collision warning, mitigation and prevention, road sign recognition, lane keeping assist, curve speed control, intersection assistance up to even autonomous driving functions.

A common clustering of ADAS has two functional dimensions, one dimension distinguishing active and passive/informative functions, the other comfort and safety [Robert Bosch GmbH, 2011b, Robert Bosch GmbH, 2011a] (see **Fig. 2**).

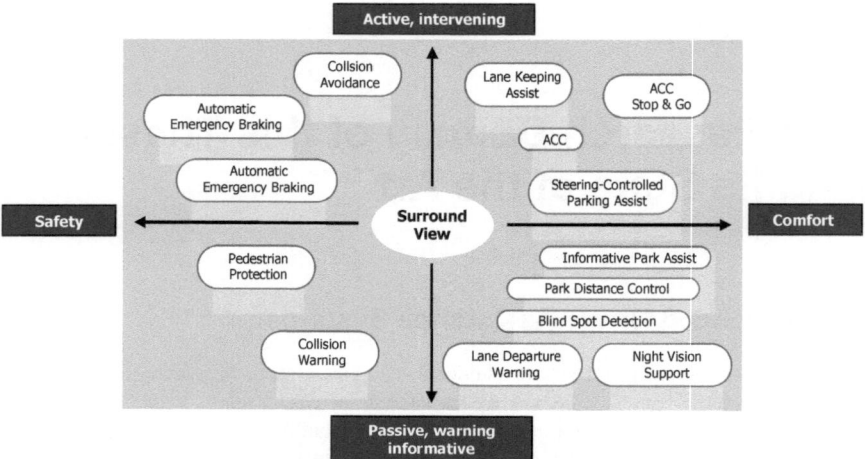

Fig. 2: Driver assistance systems (DAS) with safety and comfort functions on the basis of environmental sensors (translated from [Robert Bosch GmbH, 2011b]).

Active ADAS functions with a rather predictive nature especially benefit from this thesis, as they demand an extensive *understanding* of complex situations.

2.1.2 Sensors for Driver Assistance

Sensors may perceive information about the current state of the ego vehicle (vehicle sensors) as well as its global position (global positioning sensor) or may perceive information about the vehicle's environment (environmental sensors). A variety of sensor types for these kinds of information exists. Processing perceived sensor data is a research field on its own. Recent research especially focuses on how to combine information from different sensors.

2.1.2.1 Sensor Types

Sensor types for the ego vehicle state include wheel speed sensors, inertial sensors, steering wheel angle sensors, pedal sensors, gear sensors, pressure and temperature sensors and more. Positioning is mainly provided by GPS sensors, but new technologies such as fine positioning by wireless networks in range and positioning by landmarks are gaining interest and application.

The range of environmental sensors covers imaging sensors such as mono and stereo vision camera systems, infrared camera systems, radar sensors, laser scanners, PMD cameras and ultrasonic sensors [Robert Bosch GmbH, 2011b, Robert Bosch GmbH, 2011a]. From a wider point of view, navigation databases and communication units may be regarded as environmental sensors as well, including Vehicle-to-Vehicle (V2V) and Vehicle-to-Infrastructure (V2I) information in the vehicles perception.

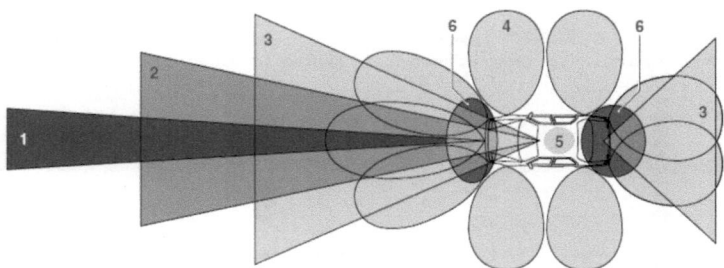

Fig. 3: Illustration of an exemplary configuration of vehicle environmental sensors and their range of perception (1 long range radar, 2 long / short range infrared sensor, 3 external video, 4 short range radar, 5 interior video, 6 ultra sonic sensors) [Reif, 2010].

The list is not meant to be complete but to provide an overview (see **Fig. 3** for an illustration [Reif, 2010]). The field of sensors is continuously and rapidly changing concerning types, variety, range, precision and capabilities in detection and classification.

2.1.2.2 Sensor Data Fusion

Data fusion serves to combine data from different sources (especially different sensors) with synergistically overlapping or complementing information to form new data or information, respectively, for extended functionality and / or better quality. In this way, improved situation estimation and improved situation responses are enabled [Steinberg et al., 1999, Robert Bosch GmbH, 2011b, Robert Bosch GmbH, 2011a].

According to the revised JDL data fusion model[4], as described and revised by [Steinberg et al., 1999, Roy, 2001, Llinas et al., 2004], data fusion may be carried out at different levels as depicted in **Fig. 4**:

- Level 0 – *Sub-object data assessment*: Pixel / signal level data association and characterization, sub-object entities (signals, features);
- Level 1 – *Object assessment*: Estimation and prediction of entities and entity states, discrete physical objects (e.g. vehicles, buildings);
- Level 2 – *Situation assessment*: Estimation and prediction of relations among entities, situation semantics;
- Level 3 – *Impact assessment*: Estimation and prediction of effects on situations of planned or estimated / predicted actions by participants;
- Level 4 – *Process refinement*: resource management with adaptive data acquisition and processing to support system objectives.

This thesis considers sensor data fusion (SDF) to be the part of data fusion that generates features and sub-features. These features are mainly objects and some basic relations that may

---

[4] originating from the Joint Directors of Laboratories Data Fusion Group (JDL DFG)

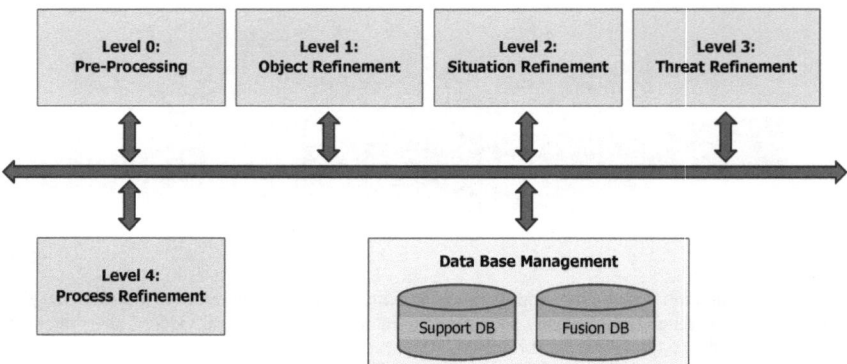

Fig. 4: The JDL data fusion model from [Polychronopoulos et al., 2006].

have attributes (sub-features) each. Basic relations are considered to be those directly deriva-
ble from sensor observations by SDF and do not need higher level logic reasoning, e.g. the
relation "is left of" to another observed moving object or the relation "is on" of a vehicle to a
road, if it can be easily obtained by sensor information. In this way SDF mainly covers levels
0 and 1 of the revised JDL data fusion model.

The JDL data fusion model has been widely adopted for automotive safety applications, e.g.
as in [Polychronopoulos et al., 2006].

Examples for the successful fusion of data from different sensor types are shown in [Liu et al.,
2008a] with the asynchronous fusion of radar and vision sensors for vehicle tracking and in
[Skutek et al., 2005] with the fusion of laser scanner and short range radar for a pre-crash system.

2.1.2.3 Types of Sensor Information

On a low level, sensors basically provide measurement signals that are processed and refined
to provide sensor features as output. Data can be fused on signal or feature level by SDF
[Dasarathy, 1997]. Within the JDL data fusion model this refers to level 0 (sub-object data
assessment). Features not further processed form *attribute features* containing values associated with a unit and potentially provided with uncertainty information such as an interval or
information about a probability distribution.

The features may be further processed for *object detection*. These *object features* relevant to
level 1 (object assessment) provide information about entities occurring in the current situation. The objects are usually tracked over time using various tracking algorithms. Detected
objects may further be *classified* as belonging to certain object types or classes. *Classification
features* may be generated within level 1 (object assessment), if only attribute features are
used. Classifying detected objects using other *object features* in addition is carried out in level
2 (situation assessment). These features are then part of *situation features*.

Generating situation features is part of a situation description as investigated in this thesis. Level 2 (situation assessment) may even generate further object and classification features and other entities such as relations not directly resulting from sensor features. Reasoning services for ontologies especially serve this task and will be extensively discussed in section 3.2 and Chapter 5 of this thesis.

Object and classification features as well as other situation features may be associated with uncertainty measures, such as existence or classification probabilities or belief intervals. This will primarily be subject-matter of section 5.4.

2.1.3 Actors for Driver Assistance

Actors for DAS systems include all types of controllable components within the vehicle to inform the driver optically, acoustically or haptically (informative HMI), to support the driver in maneuvering control (intervening HMI) or to perform autonomous actions on vehicle dynamics. Among others such actors may include displays, speakers, electric actuators and drives as well as controlled brakes or controlled steering. Details about actuators are relevant to specific DAS function realizations and not in the focus of this thesis.

2.2 Traffic Situation Description and Analysis as a Key Enabler for ADAS

The *comprehension of a traffic situation* plays a major role in driving a vehicle. Through it, perceived raw information is transformed into interpretable information. Within the driving control loop, this forms a basis for future projection, decision making and action performing, such as navigating, maneuvering and driving control. Perception, situation comprehension, projection, decision making and action performing are part of dynamic decision making introduced by [Endsley, 1995, Endsley, 2000]. This is delineated in the model of *situation awareness* (SA) depicted in **Fig. 5**. The concept of SA applies for both human decision making investigated by Endsley and for substantial advanced driver assistance systems (ADAS) as well.

The succeeding section 2.2.1 explains the concept of situation awareness by Endsley and quotes several approaches related to it before the *situation description* is integrated and discussed with respect to situation awareness in section 2.2.2.

The following section 2.2.3 deals with the main aspects and questions relevant to the generic situation description provided with this thesis. Sections 2.2.4 to 2.2.6 then quote existing approaches concerning the aspects of situation description and analysis.

2.2.1 Situation Awareness and Comprehension

Situation awareness (SA or SAW) by [Endsley, 2000] is part of the feedback loop (or control loop, respectively) of dynamic decision making. This feedback loop is shown in **Fig. 5**. The

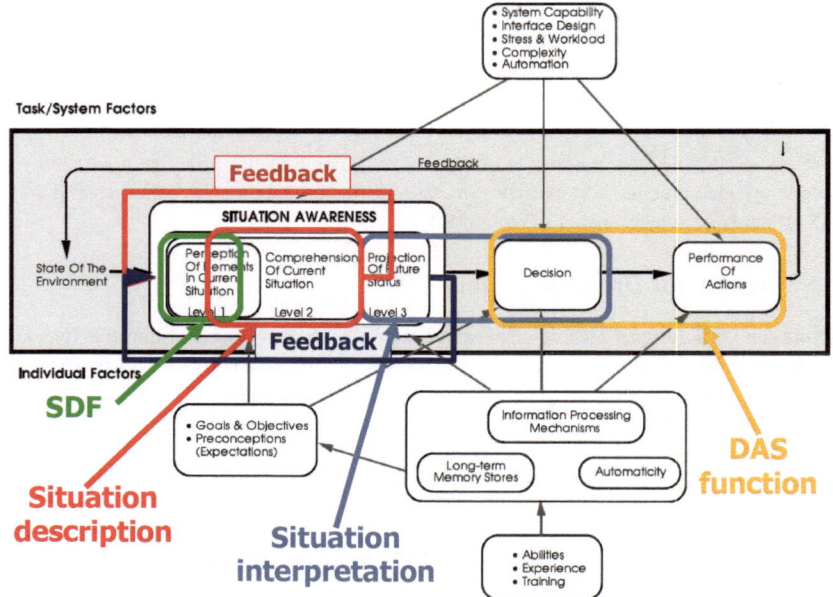

Fig. 5: Situation Awareness (SA) by Endsley within dynamic decision making [Endsley, 2000]. The components sensor data fusion, situation description, situation interpretation and DAS functions of advanced driver assistance systems are mapped to the SA model with their position in the driving control loop. Additional internal feedback loops are shown and not meant to be exhaustive.

situation awareness component captures and processes the state of environment. A succeeding component contains *decision making,* followed by a component for the *performance of actions,* which influences the state of environment completing the feedback loop.

Situation awareness by [Endsley, 2000] consists of 3 levels:

- Level 1 – *Perception of elements in current situation*: This level represents receiving raw sensor information and sensor data fusion on a low and mid level. It forms situative elements, e.g. vehicles, roads, lanes, etc., with attributes such as speed, position or size. [Endsley, 2000] especially points out the importance of choosing important information to avoid information overflow and the "odds of forming an incorrect picture".

- Level 2 – *Comprehension of current situation*: This level deals with combining, reasoning, storing and retaining perceived information. It synthesizes information from perceived information and puts it into relation. It furthermore deals with the relevance of information with respect to a person's (or in a DAS context, an application's) goals.

  Considering relevance of situation features with respect to a specific goal covers a major part of this thesis addressed in section 2.2.4, Chapter 4 and Chapter 6.

The comprehension itself with combination, reasoning, storage of and retaining the information is treated with in Chapter 5.

- Level 3 – *Projection of future status*: The third level includes forecasting future situation events and dynamics and thus *interpreting* the current situation. Anticipating future events then allows for timely decision making. It is said to be "the mark of a skilled expert" (and hence, correspondingly, a capable ADAS).

It is important at this point to distinguish the description or comprehension of a situation from an interpretation of a situation. The boundary is drawn between level 2 and level 3 SA. Level 2 reasons about current facts of the snapshot or a series of historical snapshots of a situation whereas level 3 reasons about information that cannot be told with certainty yet. In this way, different methods may be applied for situation description (e.g. knowledge-bases) and situation interpretation (e.g. probabilistic methods). This does not exclude the future application of methods that can handle both situation description and interpretation at once (e.g. probabilistic knowledge-bases).

[Roy, 2001], [Salerno, 2002] and [Hermann and Desel, 2008] compare the different levels of the JDL data fusion model from section 2.1.2.2 with the model of situation awareness and point out their commonalities and that the processes contained in both models are meant to be of parallel rather than serial nature. Especially, "the situation model should present a fused representation of the data (*situation comprehension*) and provide support for the projection needs (*situation projection*)" [Hermann and Desel, 2008]. This thesis, in particular, deals with creating a fused representation of the data for complex traffic situations.

Based on both the JDL data fusion model [Llinas et al., 2004] and the concept of situation awareness by [Endsley, 2000], [Salerno, 2002] formed a conceptual framework for a high-level architecture and functional flow addressing and detailing all levels of both models. More specific, the perception level contains object identification & tracking, historical data and a database for storage. The comprehension level contains functions such as entity extraction, filtering with learned profiles, relationship extraction, pattern matching and link analysis, model generation and model analysis. Finally, the projection layer contains the functions prediction and assessment.

The model of situation awareness by Endsley has also been mapped to unmanned vehicle situation awareness (UV SA) by [Freedman and Adams, 2007] comparing Human SA and UV SA on a high and task driven point of view.

2.2.2 Situation Description in the Driving Control Loop

The focus of this thesis is to form a *generic situation description*. With respect to ADAS described in section 2.1, the *DAS functions* cover the components *decision* and *performance of actions* of the dynamic decision making process as delineated in **Fig. 5**. The latter also contains the *actoric components* of DAS. Gathering and processing sensory information to form low level information is comprised by *sensor data fusion (SDF)* and covers the component *perception of elements in current situation.* The perception is part of *situation awareness* (SA) introduced by [Endsley, 2000] (level 1 SA), which has been explained in section 2.2.1.

Simple DAS with very specific goals such as a state of the art collision avoidance system are solely based on the selection of a target vehicle. They mainly rely on a single type of environmental sensor and a DAS function optimized on data directly acquired from the SDF. Improved and extended ADAS, gathering and combining information from several sensors and aiming to cover several DAS functions, need a generic *situation description* and *interpretation* covering extended level 2 and level 3 SA.

Situation description relevant to this thesis will be defined as the part of information processing succeeding SDF, to form a picture of the situation and provide information about it. This shall include domain specific rules and restrictions (concerning the traffic infrastructure and traffic legislature). Hence, it is aimed to further describe the evidence in a situation that is provided by sensors and preprocessed by SDF.

Situation description is not meant to include uncertain interpretations about the current situation such as predictions, behavior, uncertain intent or other possibilities. This is part of *situation interpretation* which follows downstream of situation description (please also refer to **Fig. 5** and section 2.2.1 for this partitioning). However, uncertainty, that is to be included in situation description as defined, is coming from sensors and SDF. Thus, different pictures of the world captured by sensors may have to be described in general[5].

In this way, *situation description* as defined in this thesis mainly covers level 2 SA (situation comprehension) excluding most of level 1 SA, the pure perception. *Situation interpretation* mostly covers level 3 SA, especially future situation projection, behavior description and uncertain interpretations of the captured situation and overlaps to the decision component as shown in **Fig. 5**.

2.2.3 Aspects of Situation Description Relevant to This Thesis

The aim of this thesis is to provide a generic traffic situation description capable of supplying various ADAS with relevant information about the current driving and traffic situation of the ego vehicle and its environment. With this information ADAS should be able to perform reasonable functions and actions and approach visionary goals such as injury and accident free driving, substantial assistance in arbitrary situations up to even autonomous driving.

This requires assessing more complex situations compared to state of the art assistance systems, potentially including even unknown situations demanding extensive comprehension of the vehicle environment and the current situation.

Specific requirements for the situation description are *modularity* and *hierarchy, expandability, exchangeability* and *scalability* for different variants and combinations of ADAS vehicle applications as well as *determinism* and *correctness*.

---

[5] Note, these pictures – or different worlds – do not have to be discrete but may also be fuzzy and continuous such as uncertainty about a velocity.

- Modularity: This allows for distributed, simultaneous and module based development and testing as well as for integrating only necessary modules specific to the application.

- Hierarchy: This, as applied in e.g. object oriented programming, allows for multi-level structure and inheritance of grouped functionality and properties.

- Expandability: The situation description should be able to be expanded for more expressivity, to capture more aspects of the situation or to handle extended sensor information to provide more information about the situation or to serve additional DAS functions.

- Exchangeability: In combination with modularity and expandability, exchangeability is useful to allow different target applications using different modules for different expressivity.

- Scalability: The situation description is aimed to scale from a very lean description with few information and for situations with a small number of elements up to a very expressive description with much and diverse information and for arbitrary and complex situations.

- Determinism: A specific situation shall always be described in the same manner so that a certain known input set shall result in a deterministic, foreseeable situation description output. Where desired, this enables deterministic decision making (provided certain sensor inputs) or forms a certain and reproducible basis for subsequent interpretation and uncertain decision making. Although the formulation of probabilistic reasoning may be deterministic, the outcome of a certain result of probabilistic reasoning is not. In this way, uncertain rules are excluded within the meaning of this specific requirement in the scope of this thesis. However, uncertain rules and uncertain inputs are subject to discussion in sections 3.2.5 and 5.4.

- Correctness: A situation description for DAS functions with low failure rates, especially few false positives, requires consistency of the situation context with respect to domain specific rules and constraints. For example, a vehicle cannot be on two roads at the same time and it cannot move into two different directions.

A meaningful and substantial generic traffic situation description covers:

- A delimitation of the aspects and information of the situation that are to be generically described,

- Suitable methods capable to generically describe the situation and the contained information and

- The scope and target applications of the situation description.

These items will be discussed in more detail.

Delimitation of the aspects and information of the situation that are to be generically described

A situation description has to *describe* the current situation. In this way information to describe has to be perceived, that is provided by sensor data fusion (SDF) (see subsection

2.1.2.2). Then, new information is potentially generated within the situation description, using the perceived information.

The information provided by SDF contains situation features that may be simple measures or attributes with a value and a unit up to more abstract and complex features such as objects, relations, graphs etc. that may also be generated within the situation description (see subsection 2.1.2.3). Feature data for method investigation may be generated with endurance runs and test drives, simulator studies, accident studies and computer simulations.

A core question for situation description is what situation features are generally relevant to describe the situation with respect to some target application.

As a result of research on state of the art of intelligent DAS, [Fuchs et al., 2008b] point out four sub-contexts to belong to a driving situation: the *operating environment* of the vehicle, the *driver*, the *vehicle* itself and (national) *traffic regulations*. The driver, however, will not be part of the situation description proposed with this thesis, but will be seen as part of the situation interpretation particularly involving interpretation uncertainty.

Hence, a traffic situation description has to especially describe aspects of the vehicle and its environment relevant to the situation such as traffic participants, traffic objects, environment objects and relations among them.

Additionally, a traffic situation is defined by traffic rules, infrastructure rules and further constraints defining, how objects may be related and interact. Furthermore it has to be defined, how new information may be created, based on information already contained within the situation description, e.g. by deductive rules.

Traffic rules applied in this thesis are taken from [Bundesministerium der Justiz, Juris GmbH, 2009] and [Economic Commission for Europe: Inland Transport Committee, 1993], especially concerning lane usage, speed, right-of-way, traffic lights and traffic signs.

Example rules of the "Convention on Road Traffic" in [Economic Commission for Europe: Inland Transport Committee, 1993] are:

- "Instructions conveyed by traffic light signals shall take precedence over those conveyed by road signs regulating priority."

- "Where a road comprises two or three carriageways, no driver shall take the carriageway situated on the side opposite to that appropriate to the direction of traffic."

- "(a) On two-way carriageways having four or more lanes, no driver shall take the lanes situated entirely on the half of the carriageway opposite to the side appropriate to the direction of traffic.

  (b) On two-way carriageways having three lanes, no driver shall take the lane situated at the edge of the carriageway opposite to that appropriate to the direction of traffic."

- "In States where traffic keeps to the right the driver of a vehicle shall give way, at intersections [...] to vehicles approaching from his right."

Example rules of the German StVO in [Bundesministerium der Justiz, Juris GmbH, 2009] are (translated):

- "Vehicles have to use the carriageway, out of two carriageways the one on the right."
- "The speed limit in build-up areas for all vehicles is 50 km/h, outside build-up areas [...] for passenger cars [...] 100 km/h."
- "On carriageways for both directions with four lanes indicated by lane markings, both lanes left to the driving direction are reserved for oncoming traffic; they shall not be used for overtaking. The same applies for six-lane carriageways for the three lanes left to the driving direction."
- "At intersections or forks, right-of-way has the one coming from right. This does not apply,

  1. if right-of-way is specially subject to regulation by road signs or

  2. for vehicles coming from a dirt road or forest path onto another road."
- "Traffic lights take precedence over priority rules, traffic signs regulating priority and road markings."
- "Traffic lights have the color sequence green – yellow – red – red & yellow – green."

These traffic rules are exemplary and show only a subset of rules considered in this thesis. The applied rules are logic, deterministic and unambiguous.

All information perceived from sensors or communication with external sources comes along with uncertainty (e.g. existence uncertainty, stochastic uncertainty, degree of belief etc.). Furthermore, projection to the future such as behavior prediction, which is part of situation interpretation succeeding the situation description, has inherent uncertainty.

Suitable methods capable to generically describe the situation and the contained information

A variety of methods have been tried and used for different applications and expressiveness of situation description and interpretation (as exemplary described in sections 2.2.1, 2.2.5 and 2.2.6). Popular methods[6] that have been used for both situation description and interpretation are graph like structures and databases, rule sets, decision trees, state machines (SM), neural networks (NN), different knowledge-based systems including the arising ontologies, case based reasoning and various probabilistic methods such as information theory, Bayesian networks (BN), hidden Markov models (HMM), Markov logic networks (MLN), fuzzy logic (FL) or even the arising object oriented probabilistic relational models (OPRMs). Probabilistic ontologies are not yet fully developed and manageable but gaining scientific interest. Supporting logics for e.g. spatial and temporal modeling such as Region Connection Calculus (RCC) and temporal interval logic exist. Most methods for situation interpretation and decision do not have an extensive situation description, but use an expert chosen, reduced set of input data with a very lean situation description or situation description is regarded inherent to

---

[6] Please see the subsequent subsections 2.2.4, 2.2.5 and 2.2.6 for according references.

the methods. The latter inherent description, for example, can be the distinction between situations by the inner nodes of a SM, a NN, an HMM or a BN. These inner nodes and their states, however, are dependent on the modeled network and very application specific.

Suitable methods relevant to this thesis are chosen and explained in sections 2.2.4 and 2.2.6 and their application is extensively discussed in Chapter 4, Chapter 5 and Chapter 6. This thesis uses *mutual information* (MI) based methods for calculating situation feature relevance. *Ontologies* in combination with *description logic* are used in this thesis to model situative context and to conglomerate all relevant situative information for further processing and an enhancement. A *combination of both methods*, mutual information and ontologies, is then introduced to form ontology features for relevance calculation of ontology elements.

The scope and target applications of the situation description

For this thesis and within an ADAS context, the traffic situation description shall provide information through a comprehensive interface for situation interpretation, behavior description and situation prediction leading to decision making and performing actions. The situation description provided with this thesis is aimed to be generic, modular, hierarchical, expandable, exchangeable and scalable to serve arbitrary and multiple ADAS functions. Such ADAS functions in scope were exemplary named in section 2.1.1. The situation description shall furthermore contain information or be extendable to support future functions that need knowledge about the situation.

2.2.4 Situation Feature Relevance

ADAS require a comprehensive understanding of their vehicles environment and the vehicle condition. A number of partially redundant sensors is utilized therefore to generate a variety of measurement signals. These signals or features, respectively, are then further processed, adding new features. With this high level data complexity and manifoldness, interpretation and determination of the most relevant features become unwieldy even for human experts. However, selecting relevant information is already demanded by [Endsley, 2000] concerning the perception and comprehension layers in situation awareness (compare section 2.2.1).

Few work and contributions considering situation feature selection for assistance functions and situation description are existent. [Weiser, 2010] used mutual information to determine a reduced set of input signals. With these input signals a lean Bayesian network (BN) was formed for lane change prediction. The structure of the BN was determined by expert knowledge. Input signals with nearly no relationship to lane change maneuvers were disregarded. However, only eleven signals were chosen and only two features were disregarded.

[Torkkola et al., 2004] investigated sensor selection for DAS. They used different feature selection methods to compare classification results of both the Random Forests classifier and a Naïve Bayes classifier when using different numbers of selected features. Feature selection methods used were correlation-based feature selection and the Random Forests built-in feature selection following from the number of occurrences of nodes in the trees. A significant benefit of feature selection could be shown especially with the Naïve Bayes classifier. It could also be shown that a reduced number of 32 out of all 138 features used with the Random Forests classifier produced equally accurate results as when using all 138 features.

[Schneider et al., 2008] introduce a probabilistic Bayes-based situation prediction network. It incorporates a measure for feature influence towards the situation prediction and for input influence towards features. The influences are determined with an entropy based link strength measure similar to mutual information and a partial derivative measure. The situation prediction network serves to pass deviation measures of input data through the network. However, the network has to be given in advance to calculate the intermediate influence measures. A probabilistic network for an emergency braking situation was created and simulated and results include influence levels of eleven signals that served as input to the network.

Mutual Information (MI) is a well-established method for feature selection and feature comparison [Liu et al., 2008b], for example, to support image classification methods [Bonev et al., 2008] or computer aided diagnosis [Tourassi et al., 2001]. As an advantage, it does not preliminarily need a classifier or other machine learning algorithms for relevance calculation but relevance is calculated directly from feature data series.

This thesis first presents the fundamentals (section 3.1) and subsequently the possibilities of the automated application of mutual information on sensor measurement data (Chapter 4, particularly sections 4.1, 4.3 and 4.5). It explains how features are compared with respect to different target applications and selected to crucially support the development of DAS.

Part of the situation understanding are predictions, e.g. of driving maneuvers and the vehicle trajectory, which decisions of a predictive assistance system are based on. For these predictions, machine learning methods such as *Random Forests* and *Neural Networks* are frequently used [Börger et al., 2010]. Using an optimized (sub-) set of input features can significantly improve the prediction results. Accordingly, this thesis contains the application of different methods on extensive vehicle data to evaluate the relevance of individual features in critical situations in terms of driver intent (section 4.3) and results are discussed thereon.

2.2.5 Traffic Situation Description and Interpretation Approaches

From a general point of view, situation interpretation and decision approaches build a function $\mathbf{f(g)}$, mapping a situation feature input vector $\mathbf{g}$ to one or several outputs in $\mathbf{f}$. These outputs describe current maneuvers, behaviors, desired actions, predictions etc. For the functional mapping a variety of methods has been tried and used. The mappings may be as well deterministic or nondeterministic, e.g. when modeled with probabilities, beliefs or fuzzy logic.

The mappings may be explicitly modeled, e.g. with (hierarchical) state machines (SM) [Kammel et al., 2008], decision trees [Hermann and Desel, 2008], fuzzy mapping [Hermann and Desel, 2008, Pellkofer, 2003], direct mathematical functions or rule sets [Pellkofer, 2003].

The underlying structure may be modeled and the mapping itself learned from prior data, e.g. with a given structure of networks such as Bayes networks (BN) [Huang et al., 1994, Schneider et al., 2008], Hidden Markov Models (HMM) [Meyer-Delius et al., 2008], Markov Logic

Networks (MLN) [Nienhüser et al., 2011, Stiller et al., 2007], Neural Networks (NN) [Börger et al., 2010], Object Oriented Probabilistic Relational Models (OPRM) [Schamm and Zöllner, 2011, Howard and Stumptner, 2005].

Or, structure and mapping may be learned completely, e.g. with machine learning algorithms such as Random Forests (RF) [Torkkola et al., 2004] or Support Vector Machines (SVM) to give two examples.

Most of the above listed methods are introduced theoretically in [Russell and Norvig, 2003].

The listed methods are used to interpret or predict driving maneuvers or vehicle behavior [Huang et al., 1994, Torkkola et al., 2004, Stiller et al., 2007, Hermann and Desel, 2008, Meyer-Delius et al., 2008] or to provide necessary vehicle actions [Schneider et al., 2008] in the quoted exemplary approaches. They mainly use a limited, distinct set of input features, relying on dynamic data of the ego vehicle and a few objects. In few cases, some ego vehicle based relations were used. They are designed for a specific type or set of situations, e.g. a straight road with driving paths and one or few target objects, but do not cover more complex aspects such as different types of objects, intersections, turning vehicles or traffic rule restrictions. In this way, an understanding of the situative context is lacking[7].

At the example of traffic management systems [Baumgartner et al., 2008] point out that state of the art SA (discussed in sections 2.2.1 and 2.2.2) has been realized on the lower levels with rather numeric information and methods such as BNs. They discuss a shift to higher levels in current research with objects and relations necessary for higher level SA. Consequently, they motivate ontologies to be used, which will be discussed in the following section 2.2.6. In [Baumgartner et al., 2008] ontologies are considered to be capable and context modeling to be necessary to fill the gap to describe this kind of object and relational information, but not to substitute traditional approaches.

Ontology-based situation modeling is supported by [Kokar et al., 2009], citing and discussing a variety of ontological approaches for SA in the military domain. That work points out "ontology-based computing as a paradigm on which to develop computer based situation awareness processes".

A situation description containing ontological information about objects, relations and attributes is hence aimed to serve and improve traditional situation interpretation approaches. It allows more beneficial interpretation approaches by providing extensive situative context information. In this way, an ontology-based situation description tenders an improved and extended feature input vector $\mathbf{g}$ for a situation interpretation and decision function $\mathbf{f(g)}$.

Known approaches for situation description, modeling situative context information other than with knowledge-bases or ontologies, respectively, are *Situation Aspects* by [Pellkofer and Dickmanns, 2002, Pellkofer, 2003], behavior modeling with Markov logic networks (MLNs) for cognitive automobiles by [Stiller et al., 2007], the *RoadGraph* model by [Knaup

---

[7] The only approaches using more comprehensive semantics are [Nienhüser et al., 2011, Schamm and Zöllner, 2011]. They incorporate probabilistic knowledge-based approaches, which will be mentioned in section 3.2.5.

and Homeier, 2010], *Situation Graph Trees* (SGTs) by [Arens and Nagel, 2003] and OPRMs by [Howard and Stumptner, 2005].

Behavior modeling with Situation Aspects and SGTs are very application specific, because aspects, network and tree structure have to be modeled according to the application, e.g. a certain behavior. Situation aspects themselves are linguistic variables created by a fuzzy mapping assigning a membership value in addition. Fuzzy rules are executed based on given situation aspects having a high membership value and new aspects are loaded additionally according to the executed fuzzy rules and their consequences. However, situation aspects are not related to each other and fuzzy rules are aimed to directly infer necessary behavior.

Object Oriented Probabilistic Relational Models (OPRMs) are capable of describing objects, relations and attributes to form semantics about a situation. OPRMs are frame-based and dynamically build Bayesian networks using instantiated frames. In this way, they are similar to knowledge-bases or ontologies, respectively, described in the next section 2.2.6 and section 3.2. On one hand, their advantage is the inherent handling of probabilistic information. On the other hand, they do not provide the full range of logic inference as ontologies relying on a highly formalized logic, although they may be formulated using formal logic. For large OPRM networks creating even larger Bayesian networks reasoning time may increase tremendously. Recently, OPRMs have been used for situation criticality assessment in [Schamm and Zöllner, 2011]. OPRMs together with MLNs as representatives of probabilistic knowledge-bases are discussed in more detail in section 3.2.5.

Knowledge-based approaches, especially using an ontology as motivated by [Baumgartner et al., 2008] and [Kokar et al., 2009], to model situative context and semantics of a situation are discussed in the next section 2.2.6 and the approach relevant to this thesis is introduced. The approach is compared to other existing approaches using a graph-based structure, a semantic net or an ontology as well.

2.2.6 Knowledge-Based Traffic Situation Description and Understanding

State of the art DAS directly process sensor information needed to calculate the application specific functions such as collision warning and mitigation, road sign information, lane departure warning and so on. With this functionality and current sensor technology a variety of situations is addressed effectively. To achieve visionary goals as injury and accident free driving, substantial driver assistance up to even autonomous driving, more complex and maybe even unknown situations with multiple objects are to be taken in consideration and have to be assessed [Häring and Wilhelm, 2009].

Fig. 6 shows a critical situation taken from the real world encountered during the studies on this thesis. Three vehicles are following a lead vehicle, which made a change onto the left turn lane. Consequently, all vehicles behind the lead vehicle start accelerating, expecting it to turn left. Moreover, distance between the vehicles is too small for the drivers to react on a full braking maneuver performed by the vehicle directly in front. Instead of turning left, the lead vehicle then performs a sharp right turn into a small road taking the long way round using the

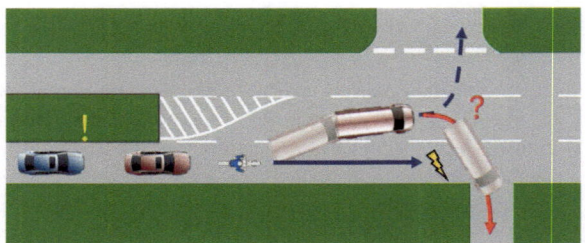

Fig. 6: Example of a critical traffic situation involving the need for a generic multi object description of traffic situations.

left turn lane. All three vehicles behind have to start an emergency braking maneuver synchronously to avoid a collision with the vehicle in front.

A state of the art collision avoidance system could react only to the vehicle directly in front, at most on the second vehicle in front. This is due to its algorithm being mostly dependent on only one selected target object [Häring and Wilhelm, 2009]. In this situation this kind of assistance would be insufficient to avoid a collision.

To assess such situations, a more general description considering the relevant objects must be provided. It should also be able to describe relations among these objects and their attributes. This gives a foundation for a situation interpretation that performs reasoning about their behavior or deviations from an expected behavior and the resulting impact as described with **Fig. 6** and sections 2.2.1 to 2.2.3. It may further be used for prediction e.g. about future positions, future maneuvers and – within the safety assistance focus of this thesis – potential risks like a collision with other traffic participants. First, an interpretation for the situation depicted in **Fig. 6** could be the prediction of the expected behavior of the lead vehicle to be turning left after moving onto the left turn lane. Then, with further situation progression, a severe deviation from the prediction would be detected when the lead vehicle starts turning right.

This validates the plausibility of the distinction between situation description and situation interpretation as defined in section 2.2.2. The correctness, quality and further use of the situation interpretation (e.g. estimated left turn or right turn, estimated future trajectories etc. in **Fig. 6**) is strongly dependent on assumptions about unknown mindsets and goals of other traffic participants and estimations of future states. It is also strongly dependent on the preliminary empirical knowledge during interpretation engineering. On the contrary, the situation description is independent to a large extent from different interpretations and stays constant. The situation description "plainly" describes the picture and potentially its history (e.g. as shown in **Fig. 6**) to serve different interpretations.

The most complex traffic situations seem to be those at intersections[8]. Their understanding is influenced by a variety of object types such as intersecting roads with lanes and markings, different kinds of road signs, allowed and forbidden paths and vehicles coming from different

---

[8] Tremendous research effort concentrates on intersection assistance. See for example PReVENT [European Commission – Information Society Technologies, 2008] and [INTERSAFE-2, 2011].

directions. Their constellation greatly influences traffic rules which apply and accordingly the assessment of allowed actions, expected behavior and impact of traffic participants among each other.

Ontologies are a foundation for *knowledge representation* and provide a formalism to structure objects, their relations and attributes and for performing *logic reasoning* with them [Nardi and Brachman, 2009]. Therefore, ontologies are well suited for modeling the kind of described multi-object traffic situations and for performing logic reasoning. Logic reasoning is able to check satisfiability of the situation ontology, to check consistency of input data and to reason about object types, relations and to e.g. apply traffic rules.

Description logic (DL) is a language for building ontologies and in most cases, depending on the dialect applied, it allows for decidable, complete and terminating algorithms. DL reasoning was extensively used by [Hummel, 2009, Hummel et al., 2007] to model geometric road infrastructure at intersections, combining the inputs of several sensors and applying rules and constraints about how infrastructure may be build up. As a consequence of the amount and detail of background knowledge modeled, reasoning on this ontology is very time-consuming and may take several hours. Basic elements of the ontology in [Hummel, 2009, Hummel et al., 2007] are used in [Vacek et al., 2007] for a case-based reasoning. These cases are generated from situations and saved in a case database. Each new case is then compared with the database and knowledge about the further situation development may be retrieved. The database is then iteratively updated with new cases. Besides reasoning time consumption, additional time and resource consumption result from the increasing size of the database with new cases. Finding good measures to compare and define cases is another challenge.

While [Hummel, 2009, Hummel et al., 2007, Vacek et al., 2007] deal with ontological modeling of infrastructure geometry and necessary elements, [Keyarsalan and Montazer, 2010] show an ontological model of a traffic flow management system for an intersection traffic light. Succeeding the ontological model, several external rules decide about traffic light cycle times and phases. These rules make use of the knowledge contained in the ontology. However, these rules do not expand the ontology by themselves and thus are not part of the ontology.

The described ontology-based methods lack extensive ontological reasoning on traffic rules with their consequences for traffic participants. Nevertheless, they form a basis for the situation description approach provided with this thesis.

[Pommerening et al., 2009] qualitatively model vehicle paths at simple intersections, using a conceptual neighborhood graph with a certain calculus to reason about positions over time at the intersection. Positions are modeled in a qualitative star like segmentation of the intersection. Their structure is restricted to be relatively simple.

[Lattner et al., 2005], [Regele, 2008] and [Knaup and Homeier, 2010] are the only known works to reason about traffic rules with ontology-like or ontological approaches. However, they do not use extended logic reasoning or inference services within the ontology.

[Regele, 2008] and [Knaup and Homeier, 2010] do not make use of a formal language for reasoning about axioms, constraints and rules like DL. Yet, their abstract graph-based, topological models are similar to the approach of this thesis to reason about traffic rules at intersections as described below. It is also based on relations about path conflicts meaning potential collisions between vehicles. In [Regele, 2008], possible conflicts at an intersection between *all* connected or crossing pairs of lanes have to be provided and only simple right-of-way rules without traffic signs or lights are considered. The RoadGraph model from [Knaup and Homeier, 2010] is capable to handle different intersection topologies and constellations of traffic signs. As a disadvantage, the RoadGraph formalization is modeled explicitly and application-specific. Thus, compared to making use of a formalized logical language like DL, it is not easily expandable, exchangeable and checkable concerning logical soundness, satisfiability and consistency.

[Lattner et al., 2005], in contrast, make use of F-Logic (a representative of FOL) as a formal language for ontological knowledge representation to store information about the traffic situation including intersections. Axioms and constraints as well as traffic rules are not included within the knowledge-base. Simple traffic rules are still applied, using queries that match situations with a potential right-of-way situation (two vehicles approaching an intersection on different roads). Hence, traffic rules are not reasoned, but certain traffic rule situations are specifically searched for. In this way, the approach is not generic with respect to traffic rules.

Similarly, Fuchs et al. [Fuchs et al., 2008b, Fuchs et al., 2008a, Fuchs, 2008] use an ontology for situation description for an overtaking assistant. Intersections are not modeled and traffic rules are mentioned, but not included with reasoning on the ontology. A sufficiently expressive rule language was not yet available and the constraint satisfaction problems (CSP), alternatively investigated, were too computationally expensive for traffic rule application.

Intersection understanding was practically used at the DARPA urban challenge in 2007 (see e.g. [Montemerlo et al., 2008, Urmson et al., 2008, Kammel et al., 2008]). Detailed prerequisite map data (road network definition file – RNDF) was provided for the urban challenge. Yet, road intersections were of a simple layout and the RNDF was extensively edited to add e.g. driving paths or critical and conflict areas.

Road maps are suitably represented by topological nets or similar graph like structures as also used in the above mentioned approaches (see e.g. RNDF [Montemerlo et al., 2008, Kammel et al., 2008], RoadGraph [Knaup and Homeier, 2010], ADAS-RP [NAVTEQ NN4D, 2011], OpenDRIVE [VIRES, 2011]). They are capable of modeling the net structure of traffic infrastructure. However, they are limited to representing it. For reasoning about e.g. traffic rules, missing infrastructure elements or traffic participants, further methods have to be used additionally. Moreover, methods for checking correctness of the provided or built up road network as well as supplementary reasoned information have to be provided.

Situations and behavior may be encoded by different active states that are responsible for the execution of situation specific functions as used in [Kammel et al., 2008] for the DARPA urban challenge. Depending on the level of detail used for the situation description with e.g. roads, lanes, lane markings, signs etc. the state machine (SM) may become very large containing a vast number of states and links among them. This complexity makes understanding and error finding very hard for scientists and engineers. [Kammel et al., 2008] used a hierarchical SM to reduce this effect.

Ontologies are powerful to model road networks as well as further traffic situation elements and provide a logic foundation for reasoning about additionally needed information as well as consistency. Whereas SMs may become very large and confusing with a rising number of considered situations, ontologies may still contain a manageable amount of logic axioms about background knowledge for situation modeling. The complexity is shifted from modeling to the automatic build up of coherent, current situation knowledge. This may furthermore go along with a higher potential to handle unknown situations with ontologies.

Consequently, this thesis proposes a lean, heavy weight ontology[9] to perform reasoning on traffic rules that participants have to obey at the example of complex road intersections (see Chapter 5). Depending on the complexity of the ontology chosen by an engineer, consistency of sensor input data may be checked to the according level of extend. The proposed ontology is structured modularly, so that expressiveness is possible to be changed easily. In this way, other ontologies e.g. from [Hummel, 2009, Hummel et al., 2007] may be connected or implemented as well. In contrast to [Regele, 2008], only necessary conflicts between vehicles are inferred within this thesis' ontology, reducing the amount of relations and reasoning effort. Furthermore reasoning is performed about right-of-way rules and traffic sign rules with or without traffic signs or lights at an intersection.

This thesis provides a sufficiently expressive ontology including reasoning as a foundation for intelligent DAS agents. These agents are empowered to ask queries to the ontology and perform their actions. Moreover, the ontology is aimed to be kept as lean as possible to speed up reasoning and allow real-time capabilities.

This thesis takes a major step towards an early application of ontologies for ADAS. Description logic based ontologies are used as a method to describe and handle the complexity of intersection structures and situations as published in [Hülsen et al., 2011b]. This thesis then makes use of it as a basis for situation interpretation and, moreover, shows real-time applicability with the implementation of the Interface and Driver Assistance Functions (IDAF) module (published in [Hülsen et al., 2011a]). It integrates the ontology into the subsequently introduced asynchronous real-time framework that utilizes a simulation software for vehicle dynamics.

2.3 Delimitation of this Thesis

In section 2.2.2 in this thesis situation description is specifically defined and differentiated from other aspects of driver assistance systems and situation awareness, especially from situation interpretation.

This thesis is not explicitly concerned with specifics of different sensor types and their functionality and it is not concerned with the lower (signal) levels of sensor data fusion, e.g. to perform object identification and tracking. Actors and DAS functions are regarded as target

---

[9] Compared to a light weight ontology for plain information storage and mostly omitted logic reasoning, a heavy weight ontology performs extensive logic reasoning on knowledge and makes use of a significant amount of axioms and rules.

applications and thus, regarded as given as well. This includes possible functions and actors in the future and is considered as the range of potential applications of the situation description provided with this thesis. Some ADAS functions are exemplary implemented for testing and for illustration of capabilities of the provided approach.

Situation interpretation as depicted in **Fig. 5** considers predictions and possible outcomes of the current situation to support opportune and advantageous decision making of ADAS functions. A variety of research work is existent and ongoing in this field. Hence, this thesis focuses on the narrowly exploited field of developing a generic *situation description*. This situation description gathers information from sensor data fusion to provide semantics of the current situation and to support situation interpretation, which is followed by the evaluation and execution of ADAS functions.

The driver is considered to be handled within situation interpretation as it is highly unobservable and ambiguous leaving large space for interpretation.

Chapter 3

Theoretical Foundations Relevant to this Thesis

This chapter gives a brief overview of the theory behind the methods applied in this thesis.

Sections 2.2.3 and 2.2.4 in Chapter 2 motivated feature selection and feature relevance estimation to be supportive of an effective and lean situation description. Mutual information was found to be a method useful to select situation features and worth further investigation.

Consequently, section 3.1 deals with feature selection methods. At first, different methods are introduced in general in section 3.1.1, before entropy and mutual information are explained in section 3.1.2 and different mutual information feature selection methods are unfolded in section 3.1.3. Finally, the Random Forest classifier is briefly presented in addition in section 3.1.4. This thesis utilizes this common classifier to compare different feature selection methods.

Furthermore, sections 2.2.3, 2.2.5 and 2.2.6 in Chapter 2 motivated knowledge-bases to be suitable for a generic situation description, especially ontologies and formal logics including logic reasoning.

Theoretical backgrounds about the domain of knowledge representation are expatiated in section 3.2. To start with, different methods for knowledge representation are told in brief in section 3.2.1, before further explaining the powerful ontologies in section 3.2.2 and their modeling with description logic (DL) as a formal language in section 3.2.3. Logic reasoning for description logic, important and extensively used in this thesis, is enclosed in section 3.2.4.

Section 2.2 pointed out, uncertainty is inherent to situation awareness, both concerning sensor inputs and situation interpretation. When using knowledge representation for situation description it is hence useful to find interfaces or aggregated methods to combine both fields. Section 3.2.5 introduces the state of research on the unification of probabilistics and knowledge representation and delimits its capabilities from the requirements of a generic situation description developed in this thesis.

3.1 Mutual Information Feature Selection

3.1.1 Feature Selection

Recently, considerable research effort has been put into the improvement of feature selection methods. These methods are mainly aiming to reduce an intractable feature set by determining the relevance of its features with respect to some goal measure [Liu et al., 2008b].

Mathematical approaches to assess the relevance of features are given in [Battiti, 1994, Peng et al., 2005]. Those approaches include methods based on mutual information (MI), chi²-test, correlation, principal component analysis (PCA), independent component analysis (ICA) as well as other related methods as e.g. variable importance related to the Random Forests (RF) classifier (see [Breiman, 2001] for RF and its variable importance measures).

Mutual information based approaches have proven to be effective and efficient on a variety of tasks. The benefit lies in its ability to determine any kind of relationship between features and moreover to be indifferent to transformations of features [Liu et al., 2008b]. Mutual information feature selection (MIFS) was first introduced in 1994 by [Battiti, 1994]. Several improvements and adaptations were given e.g. with MIFS-U[10] [Kwak and Choi, 1999], TMFS[11] [Kwak and Choi, 2002b], PWFS[12] [Kwak and Choi, 2002a], JMI[13] [Yang and Moody, 1999], MRMR[14] [Peng et al., 2005], MmD[15] [Bonev et al., 2008], NMIFS[16] and GAMIFS[17] [Estevez et al., 2009], MIREFS[18] [Liu and Hu, 2009] and comparisons to different methods are given e.g. by [Méndez et al., 2006] and [Liu et al., 2008b]. They show that no method is superior and classification results do not differ significantly in a lot of cases in [Liu et al., 2008b]. Thus potential to improvements persists.

Various applications make use of the minimum redundancy maximum relevance feature selection criterion (MRMR) [Peng et al., 2005]. Previous methodical improvements in feature selection are moreover compared to this method and it is one of the most effective and efficient methods easy to use for scientific users.

With this thesis an improvement to the MRMR-criterion is proposed by specifically calculating redundancy of features with respect to the class (section 4.2). The method will hence be named minimum class redundancy maximum relevance feature selection (MCRMR). The goal is to further improve the efficient and easy to implement MIFS-based methods.

The next section provides an introduction to mutual information followed by a section describing some of the mentioned mutual information based feature selection methods.

---

[10] Mutual Information Feature Selection with Uniformly distributed feature entropy
[11] Taguchi Method in Feature Selection
[12] Parzen Window Feature Selection
[13] Joint Mutual Information feature selection
[14] Minimum Redundancy Maximum Relevance feature selection
[15] Max-min-Dependency criterion
[16] Normalized Mutual Information Feature Selection
[17] Genetic Algorithms guided by Mutual Information for Feature Selection
[18] Mutual Information based on Renyi's Entropy Feature Selection

3.1.2 Information Theory, Entropy and Mutual Information

The entropy

$$H(X) = -\int p(X = x) \log_2 p(X = x)\, dx \tag{1}$$

determines the amount of information a random variable X contains or the uncertainty of the next random experiments outcome, respectively (see among many others e.g. [Battiti, 1994, Peng et al., 2005]).

In the discrete case the entropy is calculated with

$$H(X) = -\sum_{i=1}^{n} p(X = x_i) \log_2 p(X = x_i)\ , \tag{2}$$

accordingly. For a die roll with 6 evenly distributed die faces the information content or entropy, respectively, is given by $(6 \cdot 1/6 \cdot \log_2 6)$ bit $\approx 2{,}58$ bit.

Similarly, the conditional entropy $H(X|Y)$ is defined with

$$H(X|y) = -\int p(x|y) \log_2 p(x|y)\, dx \tag{3}$$

resulting in

$$H(X|Y) = -\iint p(y)p(x|y) \log_2 p(x|y)\, dx\, dy\ , \tag{4}$$

with according discrete formulae.

The *mutual information* (MI) is the application of the entropy on two random variables F, C and calculated as

$$I(F;C) = I(C;F) = H(C) - H(C|F)\ , \tag{5}$$

with the conditional entropy $H(C|F)$, and the closed form

$$I(F;C) = \iint p(f,c) \log_2 \frac{p(f,c)}{p(f)p(c)}\, df\, dc\ . \tag{6}$$

and

$$I(F;C) = \sum_{c \in C} \sum_{f \in F} p(f,c) \log_2 \frac{p(f,c)}{p(f)p(c)} \tag{7}$$

for the discrete case.

It indicates how much information the random variables F, C have in common or in other words the amount of information in C that can be explained by F. **Fig. 7** illustrates the concept in a set theoretical analogy.

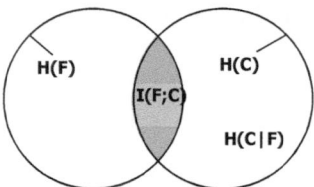

Fig. 7: Set theoretical analogy to illustrate mutual information between two random variables, e.g. a feature F and a class C.

To calculate the MI according to (5), (6) the (joint) probability distributions of the random variables are required. Histograms are taken in general. However, a sufficient number of data points and meaningful interval limits have to be provided for a good approximation.

Detailed information about entropy, MI and their application are included in [Battiti, 1994, Peng et al., 2005, Kwak and Choi, 1999, Kwak and Choi, 2002a, Yang and Moody, 1999, Bonev et al., 2008, Liu and Hu, 2009]. They also provide methods for better estimation of probability distributions and MI when simple histograms are insufficient, such as *Frasers Algorithm* [Battiti, 1994] or *Parzen Windows* [Kwak and Choi, 2002a]. [Battiti, 1994] additionally provides a formula for the approximate overestimation of the mutual information for a given discretization and a given number of data points (estimated systematic discretization error).

3.1.3 Mutual Information Feature Selection Methods

3.1.3.1 Optimal Selection Criterion

Feature selection by use of mutual information aims to optimize the set of m features $\mathbf{S}_m$, selected out of the initial set of all features $\mathbf{F}$, in a way that the joint mutual information between all the selected features together on one hand and a class variable C on the other hand is maximized (see also e.g. [Battiti, 1994]):

$$\mathbf{S}_{m,\text{opt}} = \operatorname*{argmax}_{\mathbf{S}_m \subseteq \mathbf{F}} I(\mathbf{S}_m; C) \ . \tag{8}$$

The calculation of (8) becomes intractable for large feature sets $\mathbf{F}$ and large m as well, since permutation of features to perform all MI calculations is necessary.

This can be shown with the general chain rule for joint MI (see also [Yang and Moody, 1999]):

$$I(X_1, \dots, X_{n-1}, X_n; Y) - I(X_1, \dots, X_{n-1}; Y) = I(X_n; Y | X_1, \dots, X_{n-1}) \ , \tag{9}$$

where a comma (",") separates the random variables regarded jointly concerning mutual information towards the random variable separated by the semicolon (";") as already introduced in (5).

For the 2 feature case with features X, Y this becomes

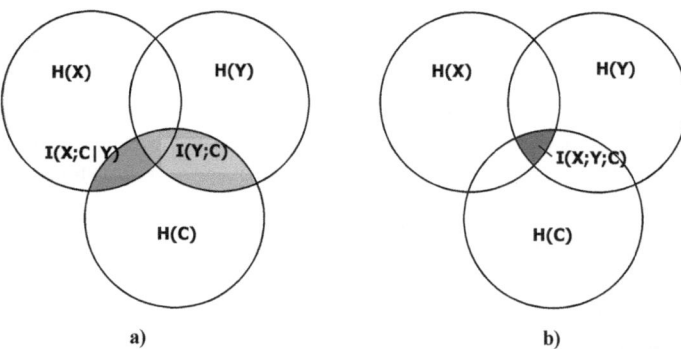

Fig. 8: **a)** Illustration of the joint mutual information for the 2 feature case. **b)** Illustration of the multivariate information between 3 features (Note this area may also be negative due to nonlinear relationships, e.g. in the case $C = X$ xor Y. For more information see [Battiti, 1994].).

$$\underset{S_2,|S_2|=2}{\operatorname{argmax}} I(\mathbf{S}_2; C) = \underset{X,Y}{\operatorname{argmax}} I(X, Y; C) \tag{10}$$

with

$$I(X, Y; C) = I(X; C) + I(Y; C|X) = I(Y; C) + I(X; C|Y)$$
$$= I(X; C) + I(Y; C) - I(X; Y; C) , \tag{11}$$

as illustrated by the light and dark grey areas in **Fig. 8 a)**. The joint MI $I(X, Y; C)$ is depicted by the complete grey area. $I(X; Y; C)$ forms the multivariate mutual information also denoted as the mutual interaction between X, Y, C (see illustrated in **Fig. 8 b)**) [Srinivasa, 2005]. It can be calculated with (9) and the general chain rule for multivariate MI (see [Batina et al., 2010]):

$$I(X_1; \dots; X_n) = I(X_1; \dots; X_{n-1}) - I(X_1; \dots; X_{n-1}|X_n) . \tag{12}$$

The goal of all further MI based feature selection methods is to find efficient approximations to (8). Due to exponentially increasing complexity carrying out an exhaustive search with all permutations of possible feature sets, most methods make use of a greedy incremental search adding one feature to the selection set at a time as shown in the next subsections.

3.1.3.2 Feature Ranking by Maximum Relevance

Given a set of random variables F_j *(features)* and a random variable C *(class)* the *maximum relevance criterion* (MaxRel) determines the MI between all features and the class separately and puts it in order [Battiti, 1994]. The incremental calculation of feature F_m at position m within the ranking is carried out with

$$F_m = \underset{F_j \in \mathbf{F} - \mathbf{S}_{m-1}}{\operatorname{argmax}} I(F_j; C) \tag{13}$$

and $\mathbf{F}$ as the initial set of features and $\mathbf{S}_{m-1}$ as the set of the $m - 1$ already selected features.

The ranking provides the following information:

- The strength or weakness of relationship between features and the class.
- The amount of information about the class a feature contains.
- The amount of mutual information the feature provides proportionally to the overall information of the function and the amount of information that is exhausted from the feature itself proportionally for the function.
- A comparison between features by the above means.

The ranking, however, provides no information about the following topics. Proposed solutions in this context are discussed later on.

- The amount of information features share among each other: For example, the features Y and Y^3 contain the same amount of information (bijective statistical mapping). If one feature is composed of several other features (feature generation) or two features are the measurement signals of two redundant sensors, they may share large amounts of information and appear directly next to each other in the resulting ranking. This will discriminate other features with new information.

  This problem can be countered by performing the ranking on features among one another. The problem is circumvented to a certain extent by further MIFS Rankings (Mutual Information Feature Selection), especially the MRMR ranking (subsection 3.1.3.5).

- The amount of information that is shared by two features with respect to the function: It is possible that only two features together actually contain information with respect to the function. An example gives the function $C = X \operatorname{xor} Y$ using uniformly distributed random variables X, Y. Let $H(X) = H(Y) = H(C) = 1$, then it follows $I(X; C) = 0$ and $I(Y; C) = 0$. However, it is $I(X, Y; C) = 1$. Accordingly, $I(X, Y; C)$ is the *Joint Mutual Information* [Yang and Moody, 1999], the information that the features X, Y provide to C together (illustrated in **Fig. 8**). In this case, the multivariate mutual information as in (11) becomes negative; it is $I(X; Y; C) = -1$ in this special example.

  In particular, this will distort the ranking and possibly make it incorrect. For example, if a variable Z exists with $0 < I(Z; C) < 1$, it will be lifted to the first position in the ranking, X or Y to the second, although X, Y together provide the full information.

  A solution to this problem is the maximum relevance ranking with the joint mutual information as in [Yang and Moody, 1999], which considers several features simultaneously. Generally, however, it can only consider as many dependencies together, as features are used in the joint mutual information. The computational effort increases exponentially with the number of features.

3.1.3.3 Extensions to Reduce Feature Redundancy

As the maximum relevance criterion does not take into account redundancy of features, further adaptations are treating this issue. The mutual information feature selection algorithm (MIFS) [Battiti, 1994] shown in (14) subtracts another term weighted by β. This term contains the sum of the MI of the currently considered feature with all already selected features. This is done in each incremental step. In that way, it accounts for the redundancy of the newly selected feature with the already selected ones.

$$F_m = \underset{F_j \in F - S_{m-1}}{\mathrm{argmax}} \left[I(F_j; C) - \beta \sum_{F_i \in S_{m-1}} I(F_i; F_j) \right] . \tag{14}$$

3.1.3.4 MIFS-U with Uniform Information Distribution

The MIFS-U algorithm [Kwak and Choi, 1999] improves selection results by extending the weight in (14) with feature dependent entropy and MI by multiplying $I(F_i; F_j)$ with $I(F_i; C)/H(F_i)$ to give a better measure for redundancy.

3.1.3.5 Minimum Redundancy Maximum Relevance Ranking (MRMR)

An improved method for the determination of important features, that contain little information at the same time with each other, comes from [Peng et al., 2005]. The *Minimum Redundancy Maximum Relevance* ranking (MRMR) adjusts the weight in (14) to average redundancy terms and thus keeps the dimension of both terms comparable within the greedy search. The incremental formula for building the ranking is

$$F_m = \underset{F_j \in F - S_{m-1}}{\mathrm{argmax}} \left[I(F_j; C) - \frac{1}{m-1} \sum_{F_i \in S_{m-1}} I(F_i; F_j) \right] . \tag{15}$$

When applying feature selection for machine learning methods, the MRMR ranking entails better results, e.g. in classification, than the maximum relevance ranking [Peng et al., 2005, Liu et al., 2008b]. Features containing new information regarding the function and little information with previously selected features are rather preferred. However, expressiveness about the comparison of features is lost, since the selection of the next feature in the ranking is strongly dependent on the previously chosen features. If one feature is excluded from the ranking, the ranking itself may change substantially.

The normalized MIFS (NMIFS) [Estevez et al., 2009] takes a similar approach by normalizing the MI term $I(F_i; F_j)$ with the minimum of the feature dependent entropies, making it $I(F_i; F_j)/\min\{H(F_i), H(F_j)\}$.

3.1.4 Random Forests, a Machine Learning Algorithm

The *Random Forests* (RF) algorithm [Breiman, 2001] is a concrete realization of machine learning. It is used to estimate functional relationships. It is currently widely used for classification tasks due to its efficient computation, stable results and good generalization capability. In this thesis it is in particular used to predict future events. On the basis of training data, machine learning algorithms generally determine a function

$$g(\mathbf{F}) = \mathbf{C} , \tag{16}$$

or a good estimation for g, mapping a matrix with feature vectors $\mathbf{F}$ onto the class vector $\mathbf{C}$ that is to be estimated. The training data consist of feature vectors $\mathbf{F}_j$ and the respective ground truth values C_j. Based on this data, an algorithm for machine learning searches for patterns in the indicator signals and generates the appropriate function $g(\mathbf{F})$.

The RF algorithm creates an ensemble of decision trees, each mapping a feature vector to the class to be estimated. Through a random selection of features and training data used for each tree slightly different results emerge. With an increasing number of trees their average output provably converges towards a stable classification and overfitting is reduced. In this way, even measurement noise and inconsistent training data may be handled.

3.2 Knowledge Representation

Knowledge representation (KR), as an area of *artificial intelligence* (AI), combines the multi-disciplinary fields of *ontologies* to define and structure things, of *logic* to enable formulation of logical inference and knowledge and *computation* to enable computer systems to perform *intelligent* tasks [Sowa, 1999].

The following sections explain aspects of the first two fields: ontologies and logic. Computation is carried out by reasoners, of which one is applied in this thesis and not handled further in detail. Knowledge representation and reasoning is furthermore described in detail in [Russell and Norvig, 2003] and [Baader et al., 2010].

3.2.1 Methods for Knowledge Representation

Common techniques for knowledge representation mainly cover formal logic, rules, frames and semantic networks [Davis et al., 1993]. They may also be combined and may overlap. They are not always clearly distinguishable[19].

Semantic networks can be a form of formal logic as in [Russell and Norvig, 2003], so that the discussions between advocates of logic and advocates of semantic networks turn out to be needless.

All KR techniques contain a family of languages to formulate knowledge.

---

[19] At this point, it shall be stated that literature in the field of knowledge representation, logics, probabilistics and their combinations does often use ambiguous terminology. Originally, ontologies are a rather philosophical mindset whereas FOL or frames can be regarded as realization languages to model ontologies. Recently, however, ontologies have been used synonymous to FOL-based formulation of knowledge. In addition, frames are strictly speaking not FOL or FOL-based, but originally a conception to model classes, attributes and relations. However, FOL is powerful to model frames. Recently, frames have been more and more regarded as FOL-based, so that frame-based concepts such as OPRMs are considered FOL-based as well.
Another ambiguity often occurs with the usage of the word reasoning. However, logic reasoning has to be distinguished from probabilistic reasoning, structural reasoning, parameter reasoning, reasoning about individuals. Probabilistic reasoning can be carried out on posterior probabilities and conditional probabilities, which are sometimes regarded as model parameters. These fields of reasoning may be combined or overlap.
For probabilistic logics and hence also first order probabilistic logics, it has to be differentiated between combined probabilistic logic reasoning and probabilistic reasoning after logic reasoning is carried out.

Semantic networks were first regarded as a graphical notation for relations between concepts (or classes) [Sowa, 1987]. It was later revised in [Sowa, 1992] to represent "any patterns of interconnected nodes and arcs" able to model and visualize logics as it is adopted by [Russell and Norvig, 2003]. In this broader view, semantic networks also serve to represent and especially graphically visualize formal logic, rules, frames as well as, more general, ontologies. Many types of semantic networks have been invented such as definitional networks, assertional networks, implicational networks, executable networks, learning networks and hybrid networks as combinations of the previous types [Sowa, 1992].

Semantic networks in the original sense may for example be modeled with the standardized conceptual graphs (CG) as well as the resource description framework (RDF). Conceptual graphs are logic-based [Sowa, 1999] whereas the RDF is similar to frames [Horrocks et al., 2003]. A frame specification declares and completely defines a class. It lacks formal logic as well as frame-based semantic networks [van Harmelen et al., 2008].

Early versions of semantic web models made use of RDF. As it lacks a formal logic resulting in some disadvantages in expressiveness, a consecutive, logic-based standard for representing ontologies was created with the web ontology language (OWL) [Horrocks et al., 2003, Antoniou and van Harmelen, 2009].

Rules as a KR technique are sets of implications. It is also referred to as logic programming with Prolog as a standardized rule language [Lloyd, 1987]. It uses predicate logic or, more specifically, horn clauses with the closed world assumption (see also open world assumption, section 3.2.4.2).

Logic-based ontologies make use of a formal logic and are shortly introduced in the next sections 3.2.2 to 3.2.4.

Recent methods are trying to combine KR methods with probabilistic reasoning such as frame based OPRMs, MLNs or probabilistic description logics such as P-SHOQ. A short introduction is provided in the last subsection 3.2.5 of this chapter.

3.2.2 Ontologies

An ontology is a formalism to represent knowledge about a field of discourse. It is built up from hierarchical concepts describing object classes. Concepts are sets of objects or instances, respectively, which can be connected by relations. Roles define types of relations. Semantics in an ontology are formed by relations between objects and axioms that can further describe concepts or roles and set constraints [Nardi and Brachman, 2009]. Concepts, role definitions and axioms are incorporated within a terminological box (TBox). Instances of concepts and roles between them are conglomerated in an assertional box (ABox). Together, ABoxes and TBoxes form a knowledge-base (KB) [Nardi and Brachman, 2009, Hummel, 2009]. **Fig. 9** shows an illustration of the basic structure of ontologies.

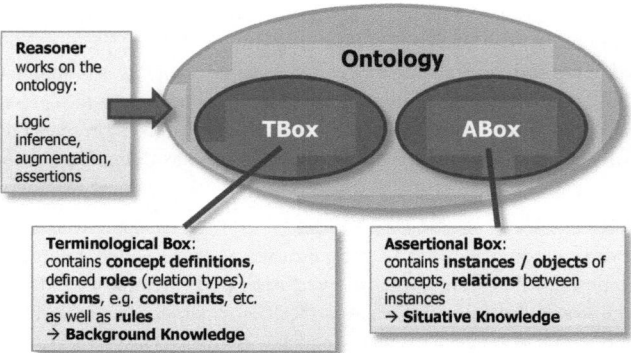

Fig. 9: Composition of ontologies containing TBoxes with background knowledge and ABoxes with situative knowledge with reasoning performed on both TBoxes and ABoxes ([Hummel, 2009] adapted and extended).

3.2.3 Description Logics

A language for representing knowledge is description logic (DL), a fragment of first-order predicate logic using only unary and binary relations [Antoniou and van Harmelen, 2009]. It provides syntax to build an ontology and semantics are described through axioms and assertions.

Various dialects exist for DL. The RacerPro reasoner [Haarslev et al., 2011] used in this thesis employs the $\mathcal{ALCQHIr+(D)}$- dialect that includes atomic negation, concept intersection, existential and universal restrictions, complex concept negation, qualified cardinal restrictions, role hierarchies, inverse roles, transitive roles and concrete domains (attribute data types). For information about DL see [Antoniou and van Harmelen, 2009, Hummel, 2009, Hummel et al., 2007, Gries et al., 2010] and for a full description consult [Baader and Nutt, 2009] or [Baader et al., 2010].

The concept and role constructors of $\mathcal{SHIQ(D)}$- (which is equal to $\mathcal{ALCQHIr+(D)}$- [Haarslev and Möller, 2001]), given in **Table 1**, underlie axiom building. Axioms themselves are then formed with the terminology given in **Table 2**. They are formed with a non-empty set Δ^I of an interpretation I. An interpretation function ($\cdot^I$) maps every concept C to a subset $C^I \subseteq \Delta^I$ of the domain of the interpretation set Δ^I and every role R to a subset $R^I \subseteq \Delta^I \times \Delta^I$. Furthermore, the constructors in **Table 1** and **Table 2** are covered by the interpretation function [Horrocks et al., 2000, Hummel et al., 2007].

Table 1: Syntax and semantics
Concept and role constructors of $\mathcal{SHIQ(D)}$- as in [Hummel et al., 2007]

Concept and role constructors		
Name	**Syntax**	**Semantics**
Top	⊤	concepts: Δ^I, roles: $\Delta^I \times \Delta^I$
Bottom	⊥	$\emptyset$
Conjunction	C ⊓ D	$C^I \cap D^I$
Disjunction	C ⊔ D	$C^I \cup D^I$
Negation	¬ C	$\Delta^I \setminus C^I$
Value restr.	∀R.C	$\{a^I \in \Delta^I | \forall b^I \in \Delta^I : (a^I, b^I) \in R^I \Rightarrow b^I \in C^I\}$
Existential restr.	∃R.C	$\{a^I \in \Delta^I | \exists b^I \in \Delta^I : (a^I, b^I) \in R^I \wedge b^I \in C^I\}$
Qualified cardinal restriction		
	$∃_{\leq n}$R.C	$\{a^I \in \Delta^I | \|\{x^I | (a^I, x^I) \in R^I, x^I \in C^I\}\| \leq n\}$
	$∃_{\geq n}$R.C	$\{a^I \in \Delta^I | \|\{x^I | (a^I, x^I) \in R^I, x^I \in C^I\}\| \geq n\}$
Role inverse	R⁻	$\{(a^I, b^I) \in \Delta^I \times \Delta^I | (b^I, a^I) \in R^I\}$

Table 2: Axiom constructors and assertions
Terminological axioms of $\mathcal{SHIQ(D)}$- as in [Hummel et al., 2007]

Assertional axioms of $\mathcal{SHIQ(D)}$- as in [Hummel et al., 2007] extended by attributes

Terminological Axioms			Assertional Axioms		
Name	**Syntax**	**Semantics**	**Name**	**Syntax**	**Semantics**
C. inclusion	C ⊑ D	$C^I \subseteq D^I$	Concept	c:C, d:D	$c^I \in C^I$, $d^I \in D^I$
C. equality	C ≡ D	$C^I = D^I$	Role	r = (c,d):R	$r^I = (c^I, d^I) \in R^I$
Role incl.	R ⊑ S	$R^I \subseteq S^I$	Equality	c = d	$c^I = d^I$
			Inequality	c ≠ d	$c^I \neq d^I$
			Attributes for concepts, roles	q = (x,a):R$_a$, x:C,R a:A	$q^I = (x^I, a^I) \in R_a^I$, $x^I \in \{C^I, R^I\}$, $a^I \in A^I$, $A^I \in \{\mathbb{N}, \mathbb{R}, \mathbb{C}, ...\}$

3.2.4 Logic Reasoning

3.2.4.1 Inference Services

Reasoners for DL provide several inference services. Inference services of the RacerPro reasoner are described in [Haarslev and Möller, 2001]. Inference services on TBoxes are subsumption, satisfiability and equivalence. *Subsumption* allows creating a hierarchical concept structure or taxonomy, respectively. *Satisfiability* checks for conflicts in concept definitions and axioms that make concepts not be instantiable. *Equivalence* between concepts holds if they always will contain the same instances.

Inference services on ABoxes include consistency, instance entailment, instance classification, instance retrieval and conjunctive queries. *Consistency* checks if the TBox allows each assertion in the ABox. *Instance entailment* checks if an assertion is satisfied by each model of the ABox. *Instance classification* finds the most specific concept to which the instance belongs. *Instance retrieval* finds all individuals of a specific concept. Eventually, *conjunctive queries* retrieve tupels of individuals for which the query conditions hold.

All standard inference services on TBoxes and ABoxes can be reduced to consistency checking. A tableau-based decision procedure is the most common implementation approach. [Horrocks et al., 2000] provide an algorithm description for a $\mathcal{SHIQ}$ ABox tableau algorithm. Simply spoken, a tableau, also called a semantic tableau or truth tree, is at the same time a decision tree and a semantic tree that allows checking satisfiability of formulas in propositional or first order logic.

Reasoning on a $\mathcal{SHIQ}$ knowledge-base is of the computational complexity EXPTIME-hard. Moreover, it is at least EXPTIME-complete, which was shown for a DL providing only conjunction, value restriction and unqualified existential restriction [Brandt, 2004] (published in [Haarslev and Möller, 2004]). [Richardson and Domingos, 2006] state logical inference w. r. t. first order logic to be NP-complete[20]. Highly optimized algorithms exist to still provide fast and efficient reasoning, even on large ABoxes, especially with the RacerPro reasoner [Haarslev and Möller, 2001]. Yet, in general, this circumstance makes a lean, effective and target oriented ontology engineering indispensable.

3.2.4.2 Open World Assumption

In contrast to binary logic where a variable that is not told to be true is always assumed false, description logic also allows assertions to be unknown. Consequently the closed world assumption (CWA) as in propositional logic is distinguished from the open world assumption (OWA). If an assertion is not stated in OWA, it is not assumed to be false or not set as a prerequisite. This characteristic of description logic inherently allows for modeling of uncertainty of perception.

3.2.4.3 ABox Augmentation Rules

To set new relation or concept assertions, rules are provided by the RacerPro reasoner [Gries et al., 2010]. They are needed to reason about relations because role chains are not provided and computationally too expensive. "In the use cases for which RacerPro is designed, we found that with a language supporting Aboxes and concrete domains as well as rules, nominals and role axioms[21] were not really strategically important" [Haarslev et al., 2011]. However, this kind of implementation using rules unfortunately does not allow checking satisfiability as it would be possible with role axioms.

Rules are formulated as Horn clauses connecting several conjunctive conditions containing several variables. Depending on the conditions the rule consequence is set. Rules are of the form

$$Q_1(\underline{Y}_1) \wedge Q_2(\underline{Y}_2) \wedge \ldots \wedge Q_n(\underline{Y}_n) \rightarrow P(\underline{X}) \tag{17}$$

---

[20] Note, that EXPTIME is suspected to be more complex than NP-complete. The proof is one of the major questions yet to answer in computational complexity theory.

[21] Role chains and other complex role axioms such as inclusion, disjointness and others

with $Q_1,...,Q_n$, P denoting concept or role names and $\underline{X}$, $\underline{Y_1},...,\underline{Y_n}$ being sets of variables. $\underline{X}$, $\underline{Y_1},...,\underline{Y_n}$ may be of arbitrary length but $\underline{X} \subseteq \underline{Y_1} \cup \cdots \cup \underline{Y_n}$ has to hold [Gries et al., 2010].

A role chain would connect two roles in the form

$$r_1 \circ r_2 \; , \tag{18}$$

which in a rule would be expressed as

$$r_1(?x, ?y) \wedge r_2(?y, ?z) \; . \tag{19}$$

Examples for rules and an alternative role chain are shown in section 5.1.2.5.

Rules are non-monotonic in reasoning so that the result depends on the order in which rules are executed. Rules especially allow the non-monotonic operator *negation-as-failure*, notated as neg(). Due to the OWA information not explicitly stated is assumed to be unknown and not to be false. In rules, the neg()-operator checks for unknown or false information and assumes this information to be false, so that a local CWA is created in the rule query. This operator explicitly draws non-monotonic results. A query containing neg(Q) yields the same result as if $\neg$ Q was stated and queried.

3.2.5 Probabilistic Logic Reasoning: State of the Art

3.2.5.1 Probabilistic Reasoning

Two types of reasoning are especially regarded to be belonging to the domain of artificial intelligence: *logical inference* and *probabilistic inference*[22] (see e.g. [Domingos et al., 2006, Russell and Norvig, 2003]). Probabilistic inference is also referred to as *Bayesian inference* being based on Bayes' theorem about conditional probabilities.

Bayesian inference may be applied to logics, within the field of artificial intelligence, especially on formal logics or frames as well. Various approaches are discussed later on in this section. Probability can be interpreted as a degree of uncertainty or a measure of belief or confidence that something is true. In this way, a certain event or statement is assigned the value 1 as probability or degree of belief. An event that does not occur with certainty or false statement is assigned the value 0. Values between 0 and 1 are considered as uncertainty. Hence, with only certain statements assigned values 0 or 1, Bayesian inference on logics becomes logical inference [Domingos et al., 2006].

Bayesian inference serves to determine posterior probabilities of random variables or hypotheses or models given some evidence (prior probabilities) and, if provided, some fixed, further conditional probabilities [Russell and Norvig, 2003]. Bayesian learning (parameter or structure learning) estimates parameters such as the conditional dependencies or the existence and direction of links between nodes.

The random variables and their conditional dependencies or conditional probabilistic relationships inherent to Bayesian inference can be represented in a graphical notation leading to *Bayesian networks* (BN) which are directed acyclic graphs. Bayesian networks are often re-

---

[22] Inference is in most cases used as a synonym to reasoning.

ferred to as causal networks, assuming the conditional dependenices are causal relationships. However, Bayesian networks with link directions all opposite are identical. Hence, a causal network is a special case of a BN. Nodes of a BN are random variables, directed links are conditional dependencies. Reasoning on a BN, given prior probabilities of evidence, leads to posterior probabilities of random variables or conditional dependencies.

A form of probabilistic graphical networks similar to Bayesian networks are *Markov random fields* (MRF), in short Markov networks, which are represented by an undirected graph [Domingos et al., 2006]. They may be cyclic and, hence, cyclic dependencies can be modeled. However, unlike BNs, MRFs cannot model induced dependencies. In MRFs, random variables are conditionally independent from all random variables they are not directly linked to. In BNs, random variables are conditionally independent from all random variables that are not their parents and not their descendants.

Probabilistic inference, in general, and hence, Bayesian network inference and Markov random field inference, are NP-hard [Zhang et al., 2006], exponential in time [Domingos et al., 2006] or #P-complete [Richardson and Domingos, 2006], respectively[23]. Sampling for approximate, fast probabilistic inference is mainly done with a Markov chain Monte Carlo (MCMC) method or derivatives, e.g. Gibbs sampling [Domingos et al., 2006]. Using sampling for probabilistic inference yields non-deterministic results.

3.2.5.2 Methods for Probabilistic Logics and Probabilistic Logic Reasoning

It has been a core interest of artificial intelligence to combine both fields of logics and probabilistics. As pointed out earlier, logic became especially relevant to formalize and reason about ontologies in the field of knowledge representation. Knowledge representation, however, often comes along with uncertainty that is expressed with probabilities or degree of belief for several uncertainty types. Hence, uncertainty is in many cases handled with probabilistic models. Further details about uncertainty are discussed in section 5.4.

This section describes some methods of combining these two fields of AI.

Probabilistic logic [Nilsson, 1986] is one the first approaches. Simply spoken, it creates a set of possible worlds of all consistent (possible) permutations of truth values of propositional logic sentences. These sentences are built from a set of Boolean variables. With assigned prior probabilities for the possible worlds, probabilities for the truth value of the sentences are calculated. Determining the consistent possible worlds is based on a semantic tree. In general, with L logical sentences, there are 2^L possible worlds subtracted by the number of inconsistent worlds, so that in the general case complexity is exponential.

Based on Nilssons probabilistic logic, advancements were developed to incorporate conditional probabilities in the semantic tree [Kane, 1989] and to extend from propositional logic to first order logic sentences [Jaumard et al., 2006]. However, besides the question what sen-

---

[23] Note, that these are not contradictory statements. A #P-(complete-)problem is at least as hard as its corresponding NP-(complete-)problem. It asks for the number of solutions of a problem rather than if there is any solution. NP-complete problems are a subset of NP-hard problems. In general, time complexity of these problems is exponential, whereas the proof of a solution can be carried out in polynomial time.

tences should be built to generate possible worlds, all worlds have to be proven to be consistent (possible) or inconsistent (impossible) which is computationally expensive. Moreover, all sentences have to be given in advance, so that logic reasoning is plainly used to derive consistency or inconsistency of worlds.

Probabilistic representations and extensions for first order logic are conglomerated by the term *first order probabilistic languages* (FOPL). A family of languages and approaches is developing within this field. Some of the recently introduced and applied methods such as object oriented probabilistic relational models (OPRMs), Markov logic networks (MLNs) and probabilistic description logic based ontologies (P-$\mathcal{SHOQ}$(D)) are briefly discussed in the remainder of this section.

OPRMs introduced in [Howard and Stumptner, 2005] and further detailed and enhanced in [Howard and Stumptner, 2006, Howard and Stumptner, 2009, Howard, 2010] extend object-oriented Bayesian networks (OOBNs) (see e.g. [Bangso and Wuillemin, 2000]). OPRMs are frame-based. Frames are defined for classes, that may be hierarchically structured, and possible relations between classes. As concretizations, instances are created from frames and form the basis for the generation of a large OOBN. Links in the generated BN are generated from the predefined sub-BNs contained in frames and external links defined between frames during frame-engineering. As a disadvantage, although OPRMs are rated among the family of FOPL, they do not provide expressiveness for logic inference apart from subsumption, such as axioms with implications, cardinal restrictions and such (compare sections 3.2.2 to 3.2.4). Logic expressiveness restricts to hierarchical classes that can be instantiated (C, $\sqsubseteq$, c:C), descriptive attributes (R_a, q:R_a) and relations between instances (R, r:R). As such, expressiveness is identical to frames that do not need FOL (see section 3.2.1). However, given a knowledge-base, OPRMs are able to perform probabilistic reasoning using well-known reasoning techniques for BNs. Early works such as [Schamm and Zöllner, 2011] show applicability in small scale. Due to their lack of logic expressiveness and reasoning capability, OPRMs are not suited for a generic situation description investigated in this thesis. Moreover, probabilistic reasoning on OPRMs would currently be too slow in medium scale, at least required for such a situation description.

MLNs, first introduced in [Richardson and Domingos, 2006] and as well belonging to FOPL, are more expressive than OPRMs. MLNs are based on MRFs (see section 3.2.5.1). Recall that nodes in MRFs represent random variables and links represent conditional dependencies between neighboring random variables. In MLNs, simple random variables are replaced by first order formulae assigned with weight. Weights denote, how strong the formula has to hold true. Formulae do not have to hold true. However, with higher weights, this is more likely in terms of probability. It is possible to set weights to infinity yielding necessarily true formulae.

Expressiveness of first-order-formulae in MLNs covers atoms such as constants, variables, functions and predicates and their combinations by negation, conjunction, disjunction, implication, equivalence, universal and existential quantification. Hence, MLNs cover most of the logic expressiveness of **Table 1**. Note, however, that MLNs do not cover qualified cardinal restrictions. Expressiveness is hence limited to $\mathcal{ALCH}$(D). In this way, MLNs cannot represent formulae such as "a car is on exactly one road".

Before creating the according MRF, the formulae are converted in their conjunctive normal form (CNF). During MRF generation, one node is assigned for each possible grounding of each predicate in the set of formulae. A grounding is a possible manifestation of an atom given the constant. Node values between 0 and 1 denote the probability that the grounding is true. Links in the network (creating the Markov blanket) are assigned between all ground atoms contained in a formula. With a given set of FOL formulae, weights can be learned from provided ground truth data. Probabilities of ground atoms are obtained by probabilistic reasoning methods on MRFs.

MLNs provide a sound method for FOPL modeling and reasoning. As a disadvantage, every possible grounding of an atom given the constants (instances) provided has to be created, regardless, if groundings and links in the MRF are inconsistent with the modeled knowledge (FOL formulae). For example, for a formula stating that, for every car (function) there exists a road (function) it is on (predicate), ground atoms with all roads and all cars and all links between them have to be created.

For large numbers of constants, atoms and large numbers and complexity of formulae, the according MRF grows very large. This extremely impedes reasonable learning of weights and tractable reasoning times for weight learning and probability calculation. With the complexity of traffic situations in the focus of this thesis, MLNs are currently intractable. However, small-scale applicability in principle has recently been shown in [Nienhüser et al., 2011] raising hope for larger-scale future applications.

A commonly used implementation for a variety of types of MLN reasoning is Alchemy (see [Domingos et al., 2006] for an algorithm description).

Two approaches for probabilistic abduction using MLNs were recently published with [Kate and Mooney, 2009] and [Gries et al., 2010].

Although MLNs are by far more logically expressive than OPRMs, they lack important axiom constructors necessary for a generic situation description. Especially, qualified cardinal restrictions are extensively used within this thesis and cannot be modeled with MLNs. Thus, MLNs have to be ruled out as well for a generic situation description relevant to this thesis at the time this thesis was created.

Another very expressive FOPL is the probabilistic extension P-$\mathcal{SHOQ}$(D) to the description logic dialect $\mathcal{SHOQ}$(D) [Giugno and Lukasiewicz, 2002]. A first implementation can be found in [Naeth and Möller, 2008]. In P-$\mathcal{SHOQ}$(D), a PTBox $\mathcal{D}$ containing conditional probabilistic constraints besides the TBox $\mathcal{T}$ as well as a PABox $\mathcal{P}$ with assertional probabilistic conditional constraints. Probabilities are modeled with a lower and an upper bound. Reasoning is performed both by a description logic reasoner and a linear program solver, since the probabilistic reasoning task is modeled as a linear program. Unfortunately, all possible, consistent logical groundings have to be reasoned out completely and, in addition, the linear program has to be solved. This results in comparably large reasoning times for even a very simple problem. In [Naeth and Möller, 2008] several seconds of reasoning time were needed for an example consisting of only 3 simple TBox formulae, 4 simple PTBox conditional constraints and 2 simple PABox conditional constraints. This makes this approach lacking in practical relevance for the near future.

Investigation of existing FOPL approaches pointed out, applicability for sufficiently expressive knowledge-bases including logical inference is, up to date, not practical. It is because of this fact that Chapter 5 mainly concentrates on pure knowledge engineering and usage of logical inference and provides a concept of some less expressive probabilistic uncertainty handling at the end.

3.3 Summary

This chapter focused on the existing theoretical foundations underlying the following chapters, which are then concerned with elaborating on the research topics of this thesis.

It introduced the concept of entropy-based mutual information and mutual information based feature selection as well as some basics of machine learning. These methods provide a basis for Chapter 4 dealing with Situation Feature Relevance on Measurement Data and its evaluations.

Secondly, it introduced concepts and methods for knowledge representation, description logics and reasoning to be extensively used in Chapter 5 dealing with Knowledge-Based Traffic Situation Description. Furthermore, probabilistic logic approaches were addressed.

Together with the results and approaches of both Chapter 4 and Chapter 5, these methods are used to form a joint approach as presented in Chapter 6. It combines mutual information feature selection and knowledge representation to deal with Relevance by Mutual Information on Ontology Features meant to facilitate increasing effectiveness and reducing complexity of the proposed Knowledge-Based Traffic Situation Description in the future.

Chapter 4

Situation Feature Relevance on Measurement Data

Predictive driver assistance systems require information about the future progress of a situation. This may be the upcoming action of the driver, the vehicle course or an event, such as a collision. Due to its complexity, this prediction is often hardly tangible for the human mind, so that machine learning methods to perform this kind of functions are increasing in popularity.

Extensive measurement data may be used to gain information concerning the future development of the situation by simply applying a time delay and may further serve as training data for machine learning algorithms and the calculation of mutual information.

Based on mutual information, the MaxRel ranking and the MRMR ranking methods are available to make use of these large amounts of measurement data. The results can help to find suitable features and to support their target-oriented generation for the development of predictive driver assistance system functions.

This chapter deals with the application of mutual information based feature selection methods as described in section 3.1.3 for feature selection on extensive vehicle measurement data. This aims to keep the state space small in the subsequent function development through a goal-oriented selection of features. The advantage of these feature selection methods is their automated applicability, especially on a large number of features or on opaque, less intuitively understandable features.

These methods can always be used when there is a fixed and known target function, which specifies a target function value for each measured data point. Examples include an offline-determined classification of maneuvers or actions to be executed, an actoric command, an offline-determined future measurement value or the current output value of an additional sensor.

Section 4.1 first explains the application of mutual information on vast automotive measurement data. Before showing a direct application, section 4.2 introduces a methodical improvement to a state-of-the-art method, which will be used in the remainder of the chapter. Subsequently, in section 4.3, results of such an application on a data set with labeled collision and de-escalation maneuvers are discussed. Section 0 shows more sophisticated methods considering more than one feature at a time and discusses their application.

Section 4.5 does not deal with the plain relevance estimation of features, but rather analyzes single features concerning their relevance of historical data. It provides investigations about typical characteristics when history is indeed of relevance. Consequently, suggestions for feature generation are provided, when typical characteristics are detected.

4.1 Relevance of Situation Features for Driver Assistance Functions

This section points out different beneficial applications of automated feature selection in the domain of driver assistance function development. It starts with the simple comparison of similar signals, goes on with the evaluation of generated features and comes to the search for features causing unwanted effects. Finally, the application of feature selection for function machine learning is briefly described.

4.1.1 Selection of Relevant Situation Features

Modern vehicles are usually equipped with various sensors. They provide a variety of measurement signals, which relate to the own vehicle, the environment, other objects, infrastructure, etc. Within this multitude of data it is useful to filter and to find relevant signals for the target function. Many of these signals may be redundant to a large extent and very similar. They can be of high importance for the function or completely irrelevant as well, without this being particularly evident.

Examples of redundant signals:

- The single wheel speeds, the engine speed, the velocity integrated over the measured acceleration and the velocity measured by environmental sensors contain similar information of different quality. In addition, it is common to have several signals of the same name available from different sources (sensors, ECUs), which are raw, preprocessed, filtered or may have been limited to required boundaries, have an offset in time or level, have been scaled or discretized and so on. It is important to find the best and necessary signals among the alternatives for an objective function.

- Radar, laser and video sensors measure the position and velocity of other objects in different ways. It is to find out, which signals of which sensor provide better information.

Whenever the comparison of various features is required, the MaxRel ranking, the MRMR ranking and the mutual information calculated for both, can be a great help and decision support. The application to real vehicle test data is presented in section 4.3.4.

4.1.2 Evaluating Generated Features

Often, new features are calculated from existing features, for example, quotients, products, integrals, gradients, relative size or frequency amplitudes and combinations of those. The generated value added or information gain, respectively, is of great interest for which the MaxRel ranking and the MRMR ranking may be used. Especially, they allow for comparison between different alternatives of generated features.

Features for the example of predictive collision avoidance include:

- Different methods of vehicle course prediction: linear, circular, parabolic, extrapolating polynomial or more complex dynamic assumptions
- To assess the criticality: use of *TTC* (Time To Collision), *TTB* (Time To Brake) or *TTR* (Time To React), *TTS* (Time To Steer), *TTA* (Time To Act) etc.

4.1.3 Finding Causes for Salient Effects

During development, unexpected results often occur and nobody has an idea of the cause, that is w.r.t functional development, the feature that is responsible for the unexpected effect. Mutual information based rankings may help finding these features. A class containing the labeled effect can be created and used as the class for the MaxRel ranking. All features with a significant cause of the effect will rank high with a high relative mutual information to the class.

4.1.4 Feature Selection for Automated Machine Learning

Machine learning methods increasingly become part of DAS development. They include *Random Forests* (RF), *Support Vector Machines* (SVMs), *Hidden Markov Models* (HMMs) and *Neural Networks* (NN). These methods learn a model from given training data with different features and the target function values. They serve to predict target functions values for new, unknown feature data.

To optimize resources (memory requirements, training time and prediction time), enable operability or in some cases even to improve the functional quality, a reduction in the number of features is useful. For this purpose the MRMR ranking from section 3.1.3.5 and the MCRMR ranking, newly introduced in section 4.2.2, are particularly suitable, as shown by the results from section 4.3.

4.2 Adaptions to Improve Accuracy

Simulations with the MRMR ranking method on data investigated in this thesis brought up suspicious errors, leading to the assumption of an inherent methodical error of at least the MRMR method, if not most of the MIFS methods introduced in section 3.1.3. These errors included the unexpected and unreasonable upper position ranking of irrelevant features, intentionally randomized probe features and even zero features (containing all zeros). Investigation of the underlying mathematical foundations of the MRMR method shows significant error potential (section 4.2.1). An improvement to the MRMR method was derived to better approximate the optimality criterion (sections 4.2.2, 4.2.4). Underpinnings of the newly introduced method are given by detailed error calculations (section 4.2.3) and simulation results on 2 data sets using different feature subsets (section 4.2.5). The newly introduced method shows tremendous benefit in classification error.

4.2.1 Related Methodical Errors

The joint MI in (8) from section 3.1.3.1 forms the optimal criterion. For the two-feature-case it can be illustrated as in **Fig. 8**. Summing over the MI between each feature and the class overestimates that optimum by the multivariate MI $I(X; Y; C)$, as shown in (10) and (11); or it underestimates that optimum in case of negative multivariate MI.

Thus by subtracting $I(X; Y)$ instead (as done by MIFS, MRMR etc.), the error of $I(X; Y|C)$ is made in the decision function as illustrated in **Fig. 10**. However, it is down-weighted by the parametric weight and normalization.

It can be observed that with the MRMR and MIFS algorithm, completely irrelevant features and even features containing all zeros or random features (noise) were ranked up high on top 10 positions. High ranking of zero and random features results from the fact that their decision term is exactly zero having no MI with other features at all, while others may have low MI with the class but high MI among other features so that their decision term becomes negative. Irrelevant features have low class MI but may have low MI with other features as well causing a similar effect like zero features.

As the MIFS-U algorithm is weighting the redundancy term with $I(F_i; C)/H(F_i)$ as given in section 3.1.3.4, it is trying to avoid this error and approximate the optimum. The approximation is good under the condition that the ratio between the multivariate MI and the MI between features is close to the ratio used by MIFS-U. This cannot be generalized.

4.2.2 Improved Criterion: Minimum Class Redundancy Maximum Relevance Ranking (MCRMR)

Consequently, following from (10) and (11), a new decision criterion for feature selection is proposed in this thesis that tries to eliminate the demonstrated error. In comparison to most MIFS-based algorithms it does not plainly take the sum of MI between the feature to select and the already selected features. Instead it considers the sum of the multivariate MI between these features AND the class $I(F_i; F_j; C)$ together. The decision criterion is consequentially called *minimum class redundancy maximum relevance criterion (MCRMR)* and is as follows:

$$F_m = \underset{F_j \in F - S_{m-1}}{\mathrm{argmax}} \left[I(F_j; C) - \sum_{F_i \in S_{m-1}} I(F_i; F_j; C) \right] \tag{20}$$

with

$$I(F_i; F_j; C) = I(F_i; F_j) - I(F_i; F_j|C) \tag{21}$$

or

$$I(F_i; F_j; C) = I(F_i; C) - I(F_i; C|F_j) . \tag{22}$$

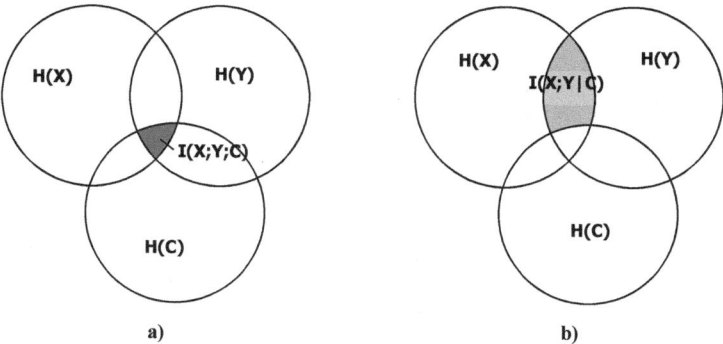

Fig. 10: Illustration of the **a)** multivariate information between 3 features (Note this area may also be negative due to nonlinear relationships, e.g. in the case $C = X$ xor Y. For more information see [Battiti, 1994].), **b)** conditional mutual information between two features given the class.

Thus, for every feature F_j, the decision term discounts the part of information included in that specific MI $I(F_j; C)$ that is explained by the already selected features. Furthermore, no weighting of the redundancy term is needed in general.

The decision to apply (21) or (22) should depend on the data and its distribution that is to be investigated. If the class C does not contain too many discrete labels and has a rather uniform distribution, then it is suitable to estimate conditional probabilities and (21) can be used. If this is not the case, as for example with the class distribution in **Fig. 16** of the collision warning data set introduced in section 4.3.3, (22) can be expected to provide better results when features can be expected to provide better conditional probabilities.

The disadvantages over optimality common with all greedy incremental methods remain as the selection of the mth feature is based on the selection result of the previous feature and does not reassess the feature set all over.

4.2.3 Error Calculations

Following calculations show the error made by the MIFS methods introduced above in section 3.1.3. It will be shown for the MRMR algorithm but can be easily transformed for the others.

The MRMR algorithm tries to maximize a function Φ_{MRMR} (from [Peng et al., 2005], slightly adapted):

$$\Phi_{\text{MRMR}}(\mathbf{S}, C) = \sum_{F_i \in \mathbf{S}} I(F_i; C) - \frac{1}{|\mathbf{S}|-1} \sum_{\substack{F_i, F_j \in \mathbf{S} \\ i<j}} I(F_i; F_j) \ . \tag{23}$$

For computational reasons, the MRMR decision criterion is derived by applying an incremental, greedy search (see section 3.1.3.5).

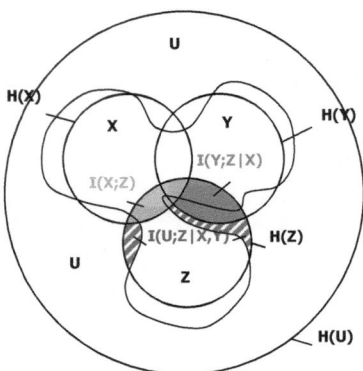

Fig. 11: Set theoretic illustration of the joint mutual information $I(X, Y, U; Z)$ (complete highlighted area) of three features X, Y, U with the class Z (different mutual information components are shown in different grey shades).

The corresponding function for the MCRMR algorithm is

$$\Phi_{\text{MCRMR}}(\mathbf{S}, C) = \sum_{F_i \in \mathbf{S}} I(F_i; C) - \sum_{\substack{F_i, F_j \in \mathbf{S} \\ i < j}} I(F_i; F_j; C) \ . \tag{24}$$

The optimal function is given in (8). For two features X, Y to select, the optimum to find is given by (11). Thus the function $\Phi_{\text{MCRMR}}(\mathbf{S}_2, C) \equiv I(X, Y; C)$ is optimal in this case. With (12) the MRMR error can be easily calculated to be

$$\Phi_{\text{MRMR}}(X, Y; C) - \Phi_{\text{MCRMR}}(X, Y; C) = -I(X; Y | C) \tag{25}$$

as shown in **Fig. 10 b)**. For more features, the error will rise accordingly by further subtracting the $I(F_i; F_j)$ terms.

An error will also arise with MCRMR for more than two features. First it will be shown for the three-feature-case with $|S| = 3$. Applying (9) and (12) to (8) it can be easily shown that

$$\begin{aligned} I(F_1, F_2, F_3; C) = {} & I(F_1; C) + I(F_2; C) + I(F_3; C) \\ & - I(F_1; F_2; C) - I(F_1; F_3; C) - I(F_2; F_3; C) + I(F_1; F_2; F_3; C) \ . \end{aligned} \tag{26}$$

For the three-feature-case the MCRMR function is

$$\begin{aligned} \Phi_{\text{MCRMR}} I(F_1, F_2, F_3; C) = {} & I(F_1; C) + I(F_2; C) + I(F_3; C) \\ & - I(F_1; F_2; C) - I(F_1; F_3; C) - I(F_2; F_3; C) \ . \end{aligned} \tag{27}$$

Thus, the error made is $-I(F_1; F_2; F_3; C)$. These relationships are illustrated in the set theoretic analogy in **Fig. 11**.

The approach can be generalized for an arbitrary number of features with $|\mathbf{S}| = n$. The optimum (8) expands to (compare [Srinivasa, 2005])

$$I(F_1, \ldots, F_n; C) = \sum_{\substack{F_i \in S}} I(F_i; C) - \sum_{\substack{F_i, F_j \in S \\ i<j}} I(F_i; F_j; C) + \sum_{\substack{F_i, F_j, F_k \in S \\ i<j<k}} I(F_i; F_j; F_k; C) - \cdots$$

$$= \sum_{\substack{\alpha=(\alpha_1,\ldots,\alpha_k) \\ \alpha_i \in \{1,\ldots,n\} \\ \alpha_1 < \cdots < \alpha_k}} (-1)^{|\alpha|-1} I(F_\alpha; C) \tag{28}$$

where $F_\alpha = \{F_{\alpha_1}; \ldots; F_{\alpha_k}\}$ and α being a multiindex.

In the same notation, the MCRMR function is built with $\alpha_i \in \{1,2\}$. Thus, the general error made by the MCRMR function results by using $\alpha_i \in \{3, \ldots, n\}$:

$$\Phi_{\text{MCRMR}}(S, C) - I(S, C) = \sum_{\substack{\alpha=(\alpha_1,\ldots,\alpha_k) \\ \alpha_i \in \{3,\ldots,n\} \\ \alpha_1 < \cdots < \alpha_k}} (-1)^{|\alpha|-1} I(F_\alpha; C) \quad . \tag{29}$$

The MCRMR function can be easily extended to a higher upper bound of α_i considering more features in the multivariate MI reducing redundancy. Notice, this always comes along with the curse of dimensionality, increasing computational cost in time consumption and numerical instabilities corresponding to the accuracy of probability distributions, especially conditional probability distributions.

The additional error with an increasing number of elements in α and the multivariate MI for a rising number of features to select is expected to decrease. On one hand the terms alternate by their sign and on the other hand, the multivariate MIs are expected to decrease as they should have less information all in common.

Error calculations show that the commonly applied MRMR method yields a potentially tremendous error in the decision criterion with respect to the optimal criterion. This error is almost completely omitted by the newly introduced MCRMR criterion. The residual error of the MCRMR methods is inherent to the limited number of two features considered simultaneously during selection. Furthermore, the error committed by all MIFS methods due to the incremental feature selection remains. Section 4.2.5 shows the benefit of the MCRMR method over the MRMR method, which may be a tendency or even be tremendous depending on the regarded feature set.

4.2.4 Variation by Weighting to Reduce Numerical Effects

The existing MIFS and related methods as introduced in (14) and (15) include a parameter to weight the redundancy discount. This is necessary to reduce the inherently large error by subtracting the complete MI between features and to yield reasonable selection results.

With the newly introduced MCRMR algorithm, weighting is not mandatory but may have some advantages. It is possible to likewise incorporate the weight ω for the sum of multivariate MIs and with $\omega = \frac{1}{m-1}$ corresponding to MRMR results in:

$$F_m = \underset{F_j \in F - S_{m-1}}{\operatorname{argmax}} \left[I(F_j; C) - \frac{1}{m-1} \sum_{F_i \in S_{m-1}} I(F_i; F_j; C) \right] . \tag{30}$$

The benefit may come for two reasons. 1) The selection result may be distorted by nonlinear effects between features e.g. nonlinear functions resulting in higher multivariate MI. By reducing the weight parameter the influence of plain relevance is increased and may thus yield suboptimal but still useful selection results. 2) The need to calculate conditional MIs underlying the multivariate MIs may have some numerical inaccuracies as a consequence. Especially with small data sets or a highly ununiformly distributed class (or features), estimation of probability densities may be inadequate. With a weight ω, these numerical effects can be reduced. As more features are selected within the ranking the multivariate MI is calculated for more features and summed up. The same applies for the numerical error, which makes the utilization of a decreasing weighting parameter according to MRMR a reasonable approach. This may reduce numerical effects but still have a good consideration of redundancy of features especially on top ranking positions.

Notice, as shown in [Brown, 2009], that (30) is equal to JMI used in [Yang and Moody, 1999] derived from plainly maximizing $\sum I(F_i, F_j; C)$. In comparison, this thesis derives (30) by proposing a weighting factor ω on (20) to reduce numerical effects with higher feature numbers. This weighting factor may be appropriately adapted for different data sets or entirely omitted.

4.2.5 Experimental Results

4.2.5.1 Methodological Approach

To show the advantage of MCRMR, an experimental study over two different data sets and applications, respectively, was carried out and compared to MRMR.

To compare feature selection results, the Random Forests classifier[24] [Breiman, 2001] (see section 3.1.4) as a machine learning method is trained on the data sets, using a variable feature number. The best ranked features according to the current number to be chosen are taken for training and testing, respectively. Cross validation could be used on one data set, the other had a predefined training and test set due to the sequential and clustered character of the data (see data description in section 4.2.5.2). For comparison, the classification test error for different numbers of features is taken.

Classification results and learning behavior of the machine learning algorithm may vary significantly for different feature sets, training / test sets and numbers of features. To reduce this effect randomized preselection of features was applied in addition to using different data sets, cross validation (on one data set) and the choice of large / small numbers of features and data points in the data set.

---

[24] A forest of 100 trees was generated for each model.

Randomized preselection of (sub-) feature sets is useful, because a single feature set may contain stochastic effects that, by chance, make a statistically inferior method preferable over another method for that specific feature set. Using randomized preselection of feature sets, applying the rankings on each of those feature sets and finally averaging over classification errors gained with the rankings, eliminates this effect.

Consequently, the random feature sets could still contain very good or completely irrelevant features for classification. Moreover, both methods were always executed on exactly the same initial random feature set. Subsequently, the currently regarded number of the best ranked features from each method was used to train and to test the Random Forests classifier with the corresponding part of the data sets.

Thus, the rankings from the feature selection algorithms were built several times over the different randomized preselections and used for learning separately. Eventually, classification results were averaged over 10 randomizations and 7-fold cross validation (where applied) indicating the benefit of the proposed method over different applications.

For MI calculation, probability densities were estimated by histograms. Continuous signals were divided in 50 equidistant intervals covering their entire signal range.

The MRMR method was used as in (15). The *MCRMR method was used as in* (30) as this slightly further reduced classification error on a higher number of selected features compared to the use of (20). Advantages in reducing numerical effects on higher numbers of features as described above are suspected to be responsible for the additional reduction in error, as discussed above.

4.2.5.2 Collision Warning Data Set

Data Description

This thesis investigates relevance of features, which are provided by vehicle and environmental sensors, to realize an automatic collision warning system for advanced driver assistance. A learned Random Forests classifier then predicts the future situation development. Details about the data set, generation of the class label and its distribution are provided in section 4.3.1, where simulations are provided that help determining feature characteristics in the underlying data. The data set is interesting, as it provides a vast amount of data points and a rather small number of features.

The overall amount of utilized data comprises 37 features with 370,000 data points. As data consists of sequential data with object scenes, data was split into a training data set of approximately 185,000 data points and a test data set of 185,000 containing only complete coherent scenes. No cross validation was therefore applied on this data set[25].

---

[25] The collision warning data set (section 4.2.5.2) was preprocessed by other researchers within the research group, formerly working in driver assistance development, so that the exact start and end points of the sequences could not be recovered. Most ego vehicle and some environmental data of a sequence are repeated as often as there are objects in the sequence. Cross validation could then lead to repetition of sequence data in the test data set that is already contained in the training data set, which would not meet the requirement of unknown test data.

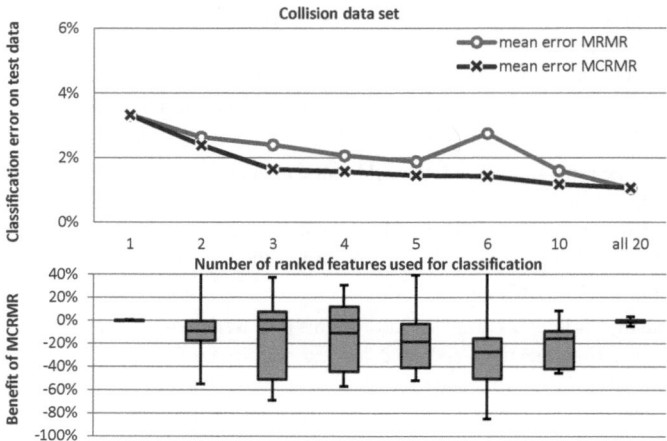

Fig. 12: Comparison of averaged classification error (top) and benefit box plots of MCRMR (as in (30)) over MRMR (bottom) on the collision warning data set for 10 times randomly taken 20 out of 37 features. (First and last value must be identical due to same feature sets. Note the nonlinear axis of the number of features.) The rising classification error originates from the augmented use of irrelevant data, when irrelevant features are added.

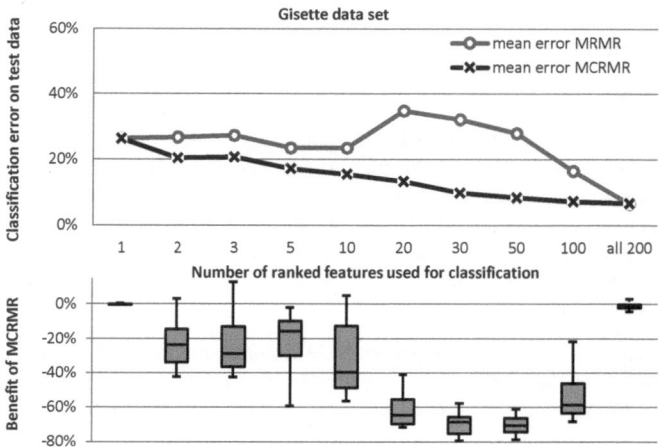

Fig. 13: Comparison of averaged classification error (top) and benefit box plots of MCRMR (as in (30)) over MRMR (bottom) on the Gisette data set for 10 times randomly taken 200 out of 5000 features. (First and last value must be identical due to same feature sets. Note the nonlinear axis of the number of features.) The rising classification error originates from the augmented use of irrelevant data, when irrelevant features are added.

To obtain an averaged comparison between MCRMR and MRMR, feature selection was not executed just once on the complete data set but on 10 randomized preselections of 20 features out of all 37 features. The data set contains no probe features with ex ante irrelevant information.

Results

Fig. 12 shows the classification results for the collision warning data set comparing the MRMR and the MCRMR method. It is plotted over the number of selected features. Classification results are averaged over 10 rankings obtained by applying the methods on the random preselections.

The proposed MCRMR method is better in average for all selected numbers of features. The average benefit in classification error over the learned numbers of features (excluding considering only the first feature and all 20 features due to methodological identity) amounts to 27.0%, with 9.7% forming the average minimum benefit and 47.7% being the average maximum benefit on this data set.

4.2.5.3 Gisette Data Set from UCI Repository

Data Description

The same experimental study was performed on a data set obtained from the UCI Machine Learning Repository [Frank and Asuncion, 2010]. The *Gisette* data set was chosen, as it provides a large number of 5000 integer features and medium sized number of 7000 data points (see [Guyon et al., 2004] for detailed information). With this application, amount and type of data and features, the data set is considerably different from the collision warning data set. It is thereby suitable to show the capability of MCRMR compared to MRMR on another application.

Half of the features are probes having no relationship with the class. It is a handwritten digit recognition problem to separate the similar digits 4 and 9. The data set was part of the NIPS 2003 feature selection challenge. With this thesis, simulations were carried out with 7-fold cross-validation creating 7 data sets with 6000 training and 1000 testing instances each. Again, to eliminate the effect of having a single set of features for feature selection, 10 random feature sets of 200 features were built to carry out both feature selection methods, MRMR and MCRMR. Therefore, considering also 7-fold cross validation, each average classification error is formed by the mean average of 70 single classification errors on 10 different feature sets.

Results

Results obtained on the Gisette data set are shown in **Fig. 13**. The benefit of MCRMR over MRMR is tremendous and even higher compared to the collision warning data set. The average benefit in classification error over the learned numbers of features (excluding considering only the first feature and all features due to methodological identity) amounts to 46.0%, with 23.9% forming the average minimum benefit and 70.5% being the average maximum benefit on this data set.

4.2.5.4 Discussion

Benefit results in **Fig. 12** and **Fig. 13** (box plots), expressed in relative reduction of classification error, show less benefit variation with an overall high and higher benefit for the Gisette data set compared to the collision warning data set. The benefit is enormous for larger feature numbers. As the Gisette data set contains probes with no relationship to the class as half of the features, MRMR will rank several of them high as described in the error discussion. MCRMR effectively avoids that.

This can also be seen in the classification error when using MRMR, rising on both data sets with a higher feature number while continually decreasing with MCRMR. As Random Forests randomly picks feature subsets during tree generation, it will also pick probes discriminating already selected better features, which impairs classification results.

The collision warning data set does not contain probes, so that this effect cannot be observed, but MCRMR still performs slightly better in average in the majority of cases on potentially useful features.

Section 4.2.2 discussed the alternative use of (21) or (22) to calculate the mutlivariate mutual information in the discount term and its effect. In the best case, the class label is nearly uniformly distributed, so that conditional probabilities can be well estimated and (21) be used. This is the case for the Gisette data set, so that the methodical benefit of the MCRMR method shows its full impact.

For the collision warning data set, conditional probabilities are harder to approximate. The class label is highly unequally distributed (for details, refer to data description and label generation in sections 4.3.1 and 4.3.3), so that the use of (21) may bias results more than the use of (22). In addition, signals are mostly continuous and probability distributions are estimated by many histogram intervals. This make a proper estimation of conditional probabilities in (22) hard as well. However, despite the difficulties in determining good estimates of conditional probabilities with the collision warning data set, MCRMR does, in average, perform at least as good or slightly better than MRMR.

4.3 Application on Measurement Data from Endurance Runs

For application in a collision warning system, the driver intention and thus the future situation progress can be estimated by the *Random Forests* algorithm (see section 3.1.4). For this purpose, the collision warning data set based on real measurement data from vehicle endurance runs is used. To show the benefit of automated feature selection and draw a comparison, the available features are prioritized by the presented mutual information based methods as well as an expert survey and randomized feature selections as references.

First, the entire process from data preparation to the application of the generated rankings is briefly introduced in subsection 4.3.1. The underlying data is described and the possible driver intentions are presented in subsections 4.3.2 and 4.3.3. Mutual information based feature

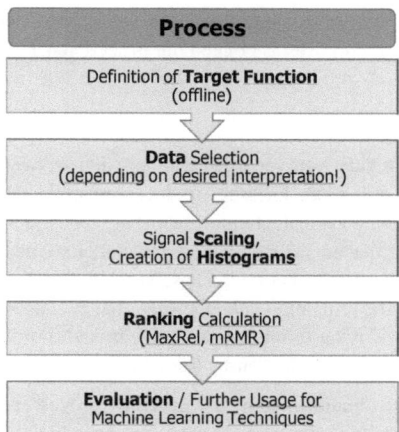

Fig. 14: Process of preparation and execution of mutual information feature selection.

rankings built with the collision warning data set are shown and discussed in subsection 4.3.4. At the end of the section, classification results of the Random Forests classifier are presented in subsection 4.3.5 and time consumption to calculate the automated feature selections are briefly compared in subsection 4.3.6.

4.3.1 The Process of Data Preparation and Ranking Creation

That the application of automated feature selection is useful to support driver assistance development was marked out in section 4.1 with respect to possible goals and will be shown at an example in the remainder of this section. Beforehand, it is important to properly apply mutual information based methods and prepare data for calculation.

Fig. 14 rouhgly shows the process from data preparation through ranking calculation to machine learning. Based on sufficient amounts of data and the feature set, which is to aimed be investigated, a class or target function has to be defined. The class label or function value must be provided for each data point. It is then important to properly select data with respect to the functional relationship that is to be determined concerning the features. This is done by data filtering. For example, if the relevance of certain features should be investigated when driving on rural roads, data has to be filtered for rural road situations. Similar filtering has to be done, if e.g. only scenes with a collision are of interest.

In the next step, histograms have to be built and priory at least the number of intervals has to be chosen. Some algorithms may be working on plain integer discretization, so that the signals have to be appropriately scaled for the required number of histogram intervals or interval size, respectively.

Succeeding data preparation, the desired rankings such as MaxRel, MRMR, MCRMR etc. can be calculated according to the formulae given in sections 3.1.3, 4.2.2 and 4.2.4. These rankings may either serve to optimize a feature set used for machine learning on the data set or directly interpreted for other purposes as pointed out in section 4.1.

4.3.2 Data Description

10.9 hours of measurement data were recorded with an Audi A8 with automatic transmission. The distance covered was 800.3 km. 1.2 hours of data were obtained with trained drivers on a test track deliberately creating critical situations, so that scenes including tight evasive steering and avoidance braking maneuvers on moving and stationary objects, as well as collisions with stationary test objects were generated. These scenes were designed primarily to test a driver assistance system. The remaining data was recorded in equal proportions in city traffic, on highways and in urban driving (common 1/3-mix). In addition to internal vehicle signals, environmental measurements of a radar sensor from the Robert Bosch GmbH were used.

Objects, that were detected simultaneously, are generally processed independently. Most ego vehicle and some environmental data of a sequence are repeated as often as there are objects in the sequence.

The measurement data amount to approximately 370,000 data points for each of 37 features. Equally to data preparation for the comparison on MCRMR and MRMR in subsection 4.2.5.2, all 370,000 data points are split into a training data set of approximately 185,000 data points and a test data set of 185,000 containing only complete coherent scenes. To calculate the mutual information, all continuous (or pseudo-continuous[26]) signals are divided into 50 equidistant histogram intervals. Discrete signals have as many histogram intervals as discrete classes.

4.3.3 Definition of the Driver Intention to be Estimated

The sought future situation progress arises from the upcoming driving maneuvers of road users. They are tantamount to the driver intention and, in an approach of the vehicle towards an object, take exactly one of the options *passing right, passing left, speed adaptation (braking)* or *collision*. In non-critical situations, the driver intention is defined as *neutral*. The maneuvers considered here are therefore also referred to as de-escalation maneuvers [Börger et al., 2010]].

To determine the driver intention, the criticality is initially determined as follows (see **Fig. 15**): [Schmidt et al., 2009] identified the estimated Time To Collision (TTC) as an important factor for human perceived criticality. Thus, time point t is considered as acutely critical, when $TTC < 3\ s$ applies and the object overlaps with the extrapolated circular course of the vehicle (lower dashed line). The entire scene is then defined as critical (dotted line), as long as the TTC is less than 3 s and the object is present. In this way, a stable off-line classification of criticality is ensured [Börger et al., 2010]].

---

[26] For example, pseudo-continuous signals are signals exhausting the full range of a discrete data type such as an integer with the range 0 to 65535.

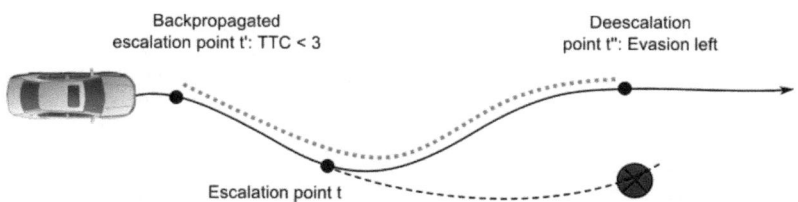

Fig. 15: Illustration of data generation by offline labeling of the criticality and driver intention. For each detected object in the situation the future situation is detected and labeled throughout the whole scene. The time t is acutely critical, and the scene is retroactively marked as critical in the range $[t', t'']$. Within this time window (thick dotted line) the driver intention is assumed to be passing left.

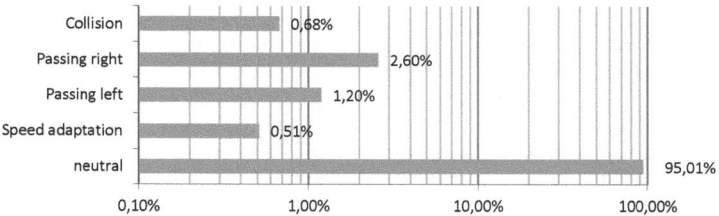

Fig. 16: Class distribution of the collision warning data set.

The driver intention is now determined at the scene ending and adopted retroactively during the critical part of the scene. For the illustrated example, it is marked as *passing left* during the period $[t', t'']$. The resulting distribution of the driver intentions in the data is shown in **Fig. 16**: Obviously, most of the obtained data is uncritical and classified as *neutral*. The critical scenes are mostly avoided by dodging, which in the class happens more frequently to the right than to the left. The reason is that, on motorways, guard rails are often detected near the left edge of the road and thus are part of the evaluation. The frequent collisions result from the test track scenes.

4.3.4 Feature Ranking Results

Table 3 shows the result of the *maximum relevance ranking* (MaxRel) on the classified measurement data. It contains the features / signals with their position in the ranking. (A feature description can be found in italics in **Table 4**. Due to nondisclosure, only a selection is listed.) Furthermore, the information content (entropy) of the de-escalation maneuver function and of all features is given. For each feature F the mutual information $I(F; C)$ with the class C is given in the next column. In addition, the proportions of mutual information w. r. t. the entropy of the class $I(F; C)/H(C)$ as well as w. r. t. the entropy of the feature $I(F; C)/H(F)$ are included. They give an impression of how much information of C is explained by F and how much information of F is used for C.

Table 3: Extract of the maximum relevance ranking (MaxRel) for de-escalating maneuvers. In addition to the features F the corresponding entropy $H(F)$, the mutual information $I(F; C)$ with the deescalation class C and the relative mutual information regarding the class and the feature entropy $I(F; C)/H(C)$ or $I(F; C)/H(F)$, respectively, are indicated. The description of the signals is given in **Table 4**.

Pos	Feature / Signal F	Entropy $H(F)$	0.3713 $I(F;C)$	$H(C)$ (De-escalation class) $I(F;C)/H(C)$	$I(F;C)/H(F)$
1	TTC	0.8120	0.2880	77.5%	35.5%
2	Object_VxRel	1.2871	0.2187	58.9%	17.0%
3	Object_DyCirc	0.8201	0.1487	40.0%	18.1%
4	Object_Dx	1.9451	0.1464	39.4%	7.5%
7	Object_Dy	1.1653	0.1401	37.7%	12.0%
9	Object_VyRel	0.7040	0.1245	33.5%	17.7%
10	Object_AxRel	1.0471	0.1064	28.7%	10.2%
14	Object_Vx	1.0665	0.0803	21.6%	7.5%
20	Object_Vy	0.3399	0.0443	11.9%	13.0%
21	EgoVx	5.4548	0.0326	8.8%	0.6%
23	EgoAx	3.2859	0.0192	5.2%	0.6%
25	WheelAngle	1.8012	0.0157	4.2%	0.9%
28	EgoAy	1.7991	0.0129	3.5%	0.7%
29	WheelAngleDt	1.1542	0.0125	3.4%	1.1%
31	GasPedal	4.2671	0.0105	2.8%	0.2%
35	GasPedalDt	2.6896	0.0056	1.5%	0.2%
36	DirIndR	0.2483	0.0011	0.3%	0.4%
37	DirIndL	0.2491	0.0009	0.3%	0.4%

Table 4: Extract of the minimum class redundancy maximum relevance ranking (MCRMR) for deescalating maneuvers. The table shows selected features / signals from the measurement data by their position in the ranking MCRMR. Note the resulting differences to the MaxRel ranking.

Pos	Feature / Signal F	Description	ΔPos
1	TTC	Object: Time To Collision	± 0
2	Object_Dy	Object: Lateral offset	− 5
3	Object_VyRel	Object: Relative lateral velocity	− 6
4	Object_Vx	Object: Longitudinal velocity	− 10
7	Object_AxRel	Object: Relative longitudinal acceleration	− 3
11	Object_DyCirc	Object: Lateral offset to circular prediction	+ 8
12	DirIndL	Indicator left (binary)	− 25
13	EgoVx	Longitudinal velocity of ego vehicle	− 8
16	Object_Vy	Object: Lateral velocity	− 4
21	EgoAx	Longitudinal acceleration of ego vehicle	− 2
25	WheelAngle	Front wheel angle of ego vehicle	± 0
26	GasPedal	Gas pedal value of ego vehicle	− 5
27	EgoAy	Lateral acceleration of ego vehicle	− 1
31	WheelAngleDt	Derivative of ego front wheel angle	+ 2
32	GasPedalDt	Derivative of gas pedal of ego vehicle	+ 1
35	DirIndR	Indicator right (binary)	− 1
36	Object_Dx	Object: longitudinal distance	+ 32
37	Object_VxRel	Object: Relative longitudinal velocity	+ 35

For example, the *TTC* has an entropy of 0.812 and the de-escalation maneuver function an entropy of 0.371. Both together share the mutual information of 0.288. This corresponds to a relative utilization of entropy of the class of $\frac{I(F;C)}{H(C)} = \frac{0.288}{0.371} = 77.5\%$. The relative utilization of entropy of the *TTC* is: $\frac{I(F;C)}{H(F)} = \frac{0.288}{0.812} = 35.5\%$.

Some interesting results of the maximum relevance ranking in **Table 3** are briefly discussed more detailed and compared to the MCRMR ranking in **Table 4**:

- The top ranked feature is by far the *TTC* (Time To Collision). Studies show that the *TTC* represents a good measure for the criticality of a situation [Schmidt et al., 2009]. Moreover and for this reason, the de-escalation maneuvers were only classified for *TTC* < 3 *s* (*neutral* otherwise), thus creating a strong functional relationship between *TTC* and the de-escalation maneuver function.

  Shortly after, at positions 2 and 4 with high mutual information as well, the features *Object_VxRel, Object_Dx,* and at position 10, the feature *Object_AxRel* follow in **Table 3**. In fact, the *TTC* is calculated from these three quantities. Here it becomes evident that the single composite feature *TTC* has a significant added value.

  In the MCRMR ranking in **Table 4**, these signals are ranked significantly lower due to their partial redundancy with the *TTC* and other early selected features (positions 4, 30 and 14 instead of 2, 4 and 10).

- The comparison of two similar features becomes evident in **Table 3** regarding the features *Object_DyCirc* and *Object_Dy*. *Object_Dy* is the lateral offset of the object to the longitudinal axis of the vehicle, *Object_DyCirc* the lateral offset to a circular prediction with the current curvature of the ego course. *Object_DyCirc* is ranked higher and is intuitively the better prediction. At high speeds, e.g. on a motorway, however, wavy lines[27] are often driven, so that *Object_Dy* might be more meaningful being the more stable feature in this case[28]. This may be the reason that the distance of both features in the ranking as well as the difference of their mutual information is not very large. As a consequence, an according speed-dependent combination should be ranked even higher[29]. In section 4.4, multiple features are considered at once for mutual information calculation and the example is revived for discussion (subsection 4.4.1, which uses the two-feature ranking in **Table 9** of 0).

- Basically, **Table 3** shows that all object data rank higher compared to the ego vehicle data, although the latter usually have a much higher signal quality. For the driver intention

---

[27] It is meant by "wavy lines" that, for example, on a straight lane the vehicle may move slightly left and right within that lane. At high speeds, these minor movements effect offsets in the lateral deviation of a circular course prediction (Object_DyCirc) that lie out of the lane itself. Hence, the lateral offset of the longitudinal vehicle axis (Object_Dy) may be a better measure at high speeds.

[28] This could be determined by filtering data and performing mutual information calculations for all data in situations with high ego vehicle speeds and low ego speeds, respectively. The comparison of both mutual information results provides direct indication, if this assumption is true.

[29] For example, a combined feature could be generated, that switches between both Object_Dy and Object_DyCirc depending on the vehicle speed.

detection and determination of possible de-escalation maneuvers, object data are therefore of great importance.

- Astonishing is the utter irrelevance of the direction indicators *DirIndL* and *DirIndR*. They do not have any influence at all to detect critical situations and the according de-escalation maneuvers. This is supported by [Schneider et al., 2008], who built a probabilistic network including influence measures for signal inputs. The learned network is used as a function to estimate a current emergency braking situation. Overall, eleven signals served as inputs. Both the direction indicator right and left have a resulting influence value of 0.

Indicators are meant to be used to indicate an upcoming lane change or turning maneuver. Obviously, there is no or hardly any dependence between activated indicators and collision warning situations. This means, critical situations may occur when the indicator is set and when it is not set as well. An example situation, one can think of, is an indicated lane change maneuver with a look over ones shoulder, while the preceding vehicle performs a sharp braking maneuver. Hence, the situation becomes critical while the indicator is set. It is easy to think of further critical situations when the indicator is not set.

In contrast to their relevance w. r. t. collision warning assistance, direction indicators are supposed have a great effect on the prediction of intended lane change maneuvers, as shown by [Weiser, 2010] who also used mutual information for some basic relevance estimation.

- In [Schneider et al., 2008], relative velocity and relative distance of the considered object to the ego vehicle (corresponding to Object_VxRel and Object_Dx in the collision warning data set) have the highest influence with influence values of 0.30 and 0.19. This is in accordance with the top ranking positions in **Table 3**. However, differences in the comparison of relevance in **Table 3** and influence of Schneider et al. occur for e.g. ego acceleration, lateral object velocity and offset as well as gas pedal position. These features locate in the midfield, but differ between rather strong or rather weak relationships.

4.3.5 Results of Driver Intention Recognition with the Random-Forest-Classifier

The determined feature rankings were now used to estimate the driver intention using the *Random Forests* classifier. Hereby, the hypothesis is tested, whether an appropriate ranking may cause better classification results or the need for fewer features.

The resulting feature rankings were evaluated as follows: Each ranking corresponds to an order of all features. A selection of a number of the first m features of one ranking is said to be better, if it yields better classification results or less classification error, respectively, than the first m features of a different ranking. Thus, for each ranking and each number m, a model was trained with the Random Forests classifier and the classification errors on the test data were compared. Hence, for each number m of selected features, there are classification errors provided for each ranking.

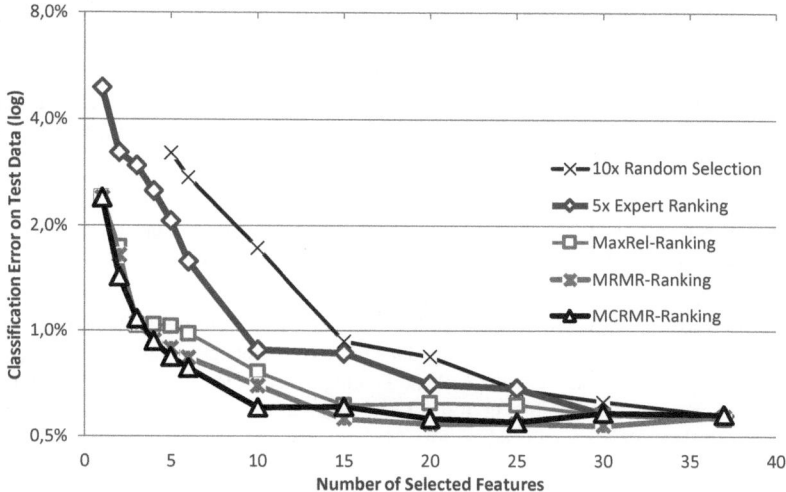

Fig. 17: Comparison of an average feature selection ranking by experts with an averaged randomized feature selection ranking and several automated mutual information feature selection rankings. The figure shows the Random Forests classification error by using the first m features of the ranking strategies each. The automated mutual information methods are all superior to expert rankings and to randomized feature selection. (Logarithmic scale)

The resulting classification errors plotted over the number m of selected features from each ranking are shown in **Fig. 17**. For $m = 37$, each classifier for each ranking uses exactly the same feature set, so that the classification error is equal for all rankings.

The following five feature rankings were evaluated:

- Random ranking (Averaged results of 10 randomized permutations)
- Expert ranking (Averaged results from five different expert rankings. The experts have been developing collision avoidance functions for years and are very familiar with the features, data recording and the function.)
- Maximum relevance (MaxRel) (see section 3.1.3.2)
- Minimum redundancy maximum relevance (MRMR) (see section 3.1.3.5)
- Minimum class redundancy maximum relevance (MCRMR) (see 4.2.2)

Consider that the maximum error when always voting the class *neutral* is at 4.99% (maximum classification error using any feature). The error in critical scenes (not *neutral*) is then about 20 times the error shown here. The random rankings could be calculated stable only with 5 features or more.

Fig. 17 shows that the MCRMR ranking delivers already very good results with few selected features selected. However, when 15 or more features are selected, the MRMR method yields slightly better results and already achieves the minimum classification error with the selection of 20 of the 37 features available. Even at 15 features, the absolute error amounts to only 0.019% above the optimum of 0.542% in classification error (3.5% relatively). At 10 features, the MCRMR ranking achieves a classification error only 0.062% above the optimum (11.4% relatively), which is remarkable, as only 27% of all features are yet considered for classification. As from three selected features already, good results are achieved, which further improve only slowly, and the classification error falls below 1.0%.

Compared to the results generated from the expert rankings, using the MCRMR ranking or the MRMR ranking yields an approximately 50% better classification error with the top 6 ranked features and yet remarkable 20% with the top 20 features used.

The maximum relevance ranking also achieved good results (approximately 10% difference in classification error). This could be due to little redundancy of most features among each other. Note, that this is also the reason, MCRMR does not have potential to significantly improve classification results compared to the MRMR ranking. This effect, caused by the nature of investigated data, is both visible in **Fig. 17** and **Fig. 12**, whereas tremendous benefit of MCRMR over MRMR is recognizable in **Fig. 13**.

4.3.6 Comparison of Time Consumption to Build Rankings

Table 5 illustrates the time needed to build different types of mutual information based feature rankings applied in this thesis.

Even with the very limited and pre-selected number of only 37 features, the experts interviewed complained about the difficulty of creating such a ranking. Relationships between the variables and the overall functional relationship are barely conceivable.

The time, required to create the ranking by each expert after explaining the task, laid between 20 and 30 minutes. For comparison, the automatic compilation of the maximum relevance ranking on a standard laptop with a 2.4GHz quad-core CPU takes about 0.6 seconds, for the MRMR ranking roughly 21 seconds and for the MCRMR ranking approximately 41 seconds. Thus, even the more complex rankings MRMR and MCRMR, which consider interdependencies between features, are already almost two dimensions faster in time, being able to use uncommonly large amounts of data. The extremely fast MaxRel ranking is a way to quickly get reasonable and interpretable results in driver assistance development where any functional relationships have to be determined.

Table 5: Time consumption of different mutual information based feature ranking approaches discussed in this thesis using a quadcore CPU with 4x 2.4 GHz.

Method	Time consumption (37 features, 370,000 data points)
Experts	~ 20 to 30 min
Max-Rel	0.59 s
MRMR	20.6 s
MCRMR	40.7 s

Overall, apart from the simple, automated handling of large data sets with many features, this comparison shows the superiority of creation time at a better performance of the investigated feature selection methods over selection by experts.

4.4 Multi Feature Rankings

4.4.1 Maximum Relevance for Multiple Features

Section 3.1.3 explained that some nonlinear relations between features and a class distort the maximum relevance ranking (MaxRel) because the ranking is built by plainly considering the mutual information between single features and the class. To some extent this issue can be met by using rankings considering redundancy between features, for example with MCRMR, MRMR or other MIFS methods from section 3.1.3.

Another way to consider nonlinear effects is to calculate the joint mutual information $I(X, Y; C) = I(X; C) + I(Y; C|X)$ for two features as given in (10) and, for the general case, given in (9). The joint mutual information is explained in more detail and moreover used for the JMI feature selection ranking by [Yang and Moody, 1999].

Maximum Relevance of two Joint Features with a Class

Yet, calculating the joint mutual information not only may be used for feature selection but also to show joint effects of features with respect to the class and in this way support function development and feature generation as explained above with the single feature rankings.

Table 6 shows an excerpt of the two feature maximum relevance ranking on the collision warning and driver intention data set from section 4.3. Besides the ranking position of feature pairs, the joint entropy, the joint mutual information with the class and accordingly, the above introduced relative mutual information values are provided. The full table (excluding undisclosed features) is provided in 0 (**Table 9**). For the 37 features, there are 703 feature pairs, including single features (regarding the feature with itself). The number of feature pairs is given by

$$n_{\text{feat.pairs}} = \sum_{k=1}^{n_{\text{feat}}} k = \frac{n_{\text{feat}}(n_{\text{feat}} + 1)}{2} \ . \tag{31}$$

Obviously, all pairs of features are ranked higher than one single feature contained in the feature pair, because $I(Y; C|X) \geq 0$ and thus $I(X, Y; C) \geq I(X; C)$. For example, the single TTC ranks on position 39. All other 36 features paired with the TTC are ranked higher. The TTC (actually $I(TTC, TTC; C)$ was calculated) has a mutual information of 0.2880, equal to $I(TTC; C)$ in the MaxRel ranking in **Table 3**. All other pairs have higher joint mutual information of up to 0.3272.

Table 6: Extract of the maximum relevance ranking with 2 joint features for de-escalating maneuvers (missing positions are due to nondisclosure). The full ranking containing all disclosed feature pairs is provided with **Table 9** in 0.

Pos	Features / Signals F,G	Entropy $H(F)+H(G\vert F)$	0.3713 $I(F,G;C)$	$H(C)$ (De-escalation class) $I(F,G;C)/H(C)$	$I(F,G;C)/H(F,G)$
1	Object_DyCirc , TTC	1,4707	0,3272	88,12%	22,25%
2	Object_Dy , TTC	1,8088	0,3233	87,05%	17,87%
3	Object_Dx , TTC	2,5172	0,3216	86,60%	12,78%
5	Object_VxRel , TTC	1,7918	0,3165	85,24%	17,67%
8	Object_Dx , Object_VxRel	2,7024	0,3121	84,06%	11,55%
10	EgoVx , TTC	6,2297	0,3112	83,80%	4,99%
12	Object_Vx , TTC	1,7846	0,3099	83,45%	17,36%
13	Object_VyRel , TTC	1,3637	0,3095	83,35%	22,70%
15	GasPedal , TTC	5,0623	0,3056	82,29%	6,04%
18	TTC , WheelAngle	2,5999	0,3035	81,74%	11,68%
19	EgoAx , TTC	4,0768	0,3034	81,69%	7,44%
22	Object_AxRel , TTC	1,7077	0,3012	81,13%	17,64%
23	EgoAy , TTC	2,6021	0,3010	81,05%	11,57%
25	GasPedalDt , TTC	3,4769	0,3003	80,87%	8,64%
27	TTC , WheelAngleDt	1,9526	0,2992	80,58%	15,32%
28	Object_Vy , TTC	1,1082	0,2986	80,40%	26,94%
37	DirIndL , TTC	1,0601	0,2893	77,91%	27,29%
38	DirIndR , TTC	1,0589	0,2890	77,84%	27,29%
39	TTC	0,8120	0,2880	77,55%	35,46%
42	Object_DyCirc , Object_VxRel	1,8137	0,2582	69,52%	14,23%
44	Object_Dy , Object_VxRel	2,1046	0,2574	69,33%	12,23%
46	GasPedal , Object_VxRel	5,4991	0,2550	68,66%	4,64%
48	Object_Vx , Object_VxRel	1,9809	0,2495	67,19%	12,60%
49	EgoVx , Object_VxRel	6,4228	0,2495	67,19%	3,88%
…	…	…	…	…	…

Notably, the only feature pair[30] without the *TTC* having more relevance than the *TTC* itself is the pair (*Object_Dx , Object_VxRel*). On one hand, this pair contains a lot of information of the *TTC*, since the *TTC* is calculated of mainly both features in the feature pair and the relative acceleration to the object in addition. On the other hand, both features contain additional relevant information to the class that cannot be explained by the *TTC*.

The ranking shows another interesting effect, that is worth to be automatically calculated. In the usual case, it should be $I(X,Y;C) \leq I(X;C) + I(Y;C)$. This is the case illustrated earlier in **Fig. 8** on page 31, where the complete grey area $I(X,Y;C)$ is calculated and in the sum $I(X;C) + I(Y;C)$ the multivariate information $I(X;Y;C) \geq 0$ is contained twice. As an example, most pairs may serve. The first pair has a joint mutual information $I(Object\_DyCirc, TTC; C) = 0.3272$, whereas the sum $I(Object\_DyCirc; C) + I(TTC; C) = 0.1487 + 0.2880 = 0.4367$. Hence, the multivariate MI is $I(Object\_DyCirc; TTC; C) = 0.1095$. That is the information contained in both features together with the class.

However, a different effect can be observed for certain feature pairs. Let us consider the pair (*GasPedal , TTC*). It has a joint mutual information $I(GasPedal, TTC; C) = 0.3056$, whereas the sum $I(GasPedal; C) + I(TTC; C) = 0.0105 + 0.2880 = 0.2985$. Hence, the multivariate MI is $I(GasPedal; TTC; C) = -0.0071$. It is negative. That is, information is not overes-

---

[30] together with another undisclosed feature pair

timated with respect to the class, but new information is actually added, when both features are considered at a time.

This effect is even bigger for the pair ($GasPedal$, $GasPedalDt$), with a joint MI of $I(GasPedal, GasPedalDt; C) = 0.0388$ and a multivariate MI of $I(GasPedal; GasPedalDt; C) = -0.0227$. The absolute multivariate MI is not much bigger than in the previous example. Yet, this case is special, as the single MIs and even their sum $I(GasPedal; C) + I(GasPedalDt; C) = 0.0105 + 0.0056 = 0.0161$ are all lower than the absolute value of the multivariate MI. In this way, these features generate useful information only if they are considered together.

To investigate such effects, it is recommended to build a ranking calculated with the multivariate MIs between feature pairs and sorting by lowest value (lowest negative value first).

Back in section 4.3.4, it was announced to revive the example discussion about the combination of features $Object\_DyCirc$ and $Object\_Dy$. These two features were supposed to have significantly higher benefit, if they are considered together in a combined feature. This should be visible in the two-feature-ranking (full ranking in **Table 9** of 0). Both single features have a MI with the class of 0.1487 and 0.1401, or 40.0% and 37.7% relative MI, respectively. In **Table 9**, the joint MI amounts to 0.1775, that is 47.8% of the class entropy. Hence, the combined consideration of both features adds roughly 8% or 10%, respectively. Remarkably, only the 20 highest ranked single features have a higher relative MI w. r. t. the class than this relative gain. The conclusion can be drawn, that a combined feature of both $Object\_DyCirc$ and $Object\_Dy$ yields significant benefit.

Maximum Relevance of three Joint Features with a Class

Considering three features at a time the formula for the joint mutual information derived from (9) is

$$I(X, Y, U; Z) = I(X; Z) + I(Y; Z|X) + I(U; Z|X, Y) \ . \tag{32}$$

Corresponding to the two feature joint mutual information calculations, a three feature maximum relevance ranking is possible. An illustration for the joint mutual information of three features X, Y, U with a class Z is shown in **Fig. 18 a)**.

Unfortunately, the currently available amounts of data – or, more precisely, the amount of accessible memory for calculation – do not allow sufficiently accurate estimation of conditional probabilities, especially conditioned over two features. Thereby, application of this method has to wait for these conditions to be fulfilled.

The attempt to apply this method on the collision warning data set did not produce accurate and usable results and classification errors compared to those in **Fig. 17** where poor compared to the MCRMR and MRMR methods.

Applying MIFS on Multiple Functions / Classes

Mutual information calculations can as well be calculated for multiple classes at a time, e.g. for finding a suitable feature set for several driver assistance functions together. The following formula, derived by applying (9) twice, may be used:

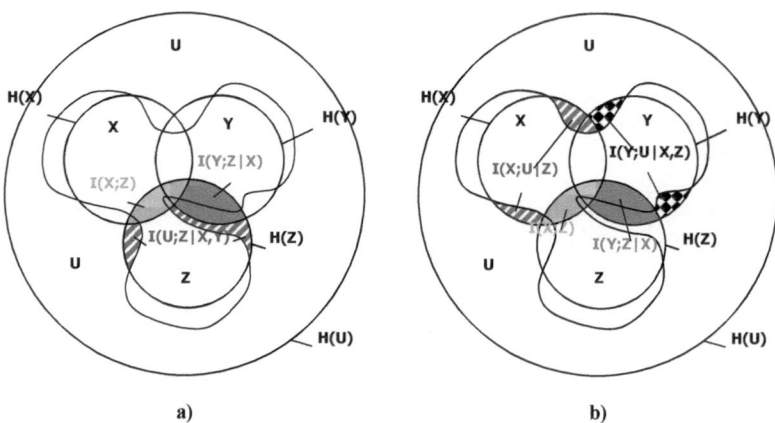

Fig. 18: a) Set theoretic illustration of the joint mutual information of 3 features X, Y, U with the class Z. **b)** Set theoretic illustration of the joint mutual information of two features X, Y with the two classes U, Z.

$$I(X, Y; U, Z) = I(X; Z) + I(Y; Z|X) + I(X; U|Z) + I(Y; U|X, Z) \ . \tag{33}$$

The mutual information calculated with (33) is illustrated in **Fig. 18 b)**. Application of this method has to wait just as the three feature joint mutual information, as sufficient amounts of data and resources for calculation have to be available first.

4.4.2 Reducing Redundancy among Pairs of Features

A reasonable consequence of building a two feature maximum relevance ranking is performing feature selection considering redundancy of feature pairs as well. A transformation of the MRMR ranking toward a 2F-MRMR ranking considering feature pairs could be

$$(F_m, F_{m+1}) = \underset{F_i, F_j \in \mathbf{F} - \mathbf{S}_{m-1}}{\mathrm{argmax}} \left[I\big((F_i, F_j); C\big) - \frac{1}{m-1} \sum_{F_k, F_l \in \mathbf{S}_{m-1}} I\big((F_i, F_j); (F_k, F_l)\big) \right] \tag{34}$$

and in further detail

$$I\big((F_i, F_j); (F_k, F_l)\big) = I(F_i; F_k) + I(F_j; F_k|F_i) + I(F_i; F_l|F_k) + I(F_j; F_l|F_i, F_k) \ . \tag{35}$$

Correspondingly, a 2F-MCRMR for feature pairs can be formalized as

$$(F_m, F_{m+1}) = \underset{F_i, F_j \in \mathbf{F} - \mathbf{S}_{m-1}}{\mathrm{argmax}} \left[I\big((F_i, F_j); C\big) - \sum_{F_k, F_l \in \mathbf{S}_{m-1}} I\big((F_i, F_j); (F_k, F_l); C\big) \right] \tag{36}$$

with

$$I\big((F_i, F_j); (F_k, F_l); C\big) = I\big((F_i, F_j); (F_k, F_l)\big) - I\big((F_i, F_j); (F_k, F_l)|C\big) \tag{37}$$

and

$$I\left((F_i, F_j); (F_k, F_l)|C\right)$$
$$= I(F_i; F_k|C) + I\left(F_j; F_k|F_i, C\right) + I(F_i; F_l|F_k, C) + I\left(F_j; F_l|F_i, F_k, C\right). \tag{38}$$

However, running the Random Forests classifier over the 2F-MRMR and 2F-MCRMR rankings does not result in superior results over using MRMR, MCRMR or in some cases even MaxRel rankings. The reason is again assumed to be in the inaccurate calculation of the conditional mutual information values in (35) and (38). Whereas the conditional mutual information in (21) or (22) to calculate the MCRMR ranking is supposed to be sufficiently accurate for most data sets, calculation of the conditional mutual information given one or even several features with many histogram intervals for probability distribution estimation almost certainly results in inaccurate conditional probabilities. This can only be met by providing enormous amounts of data and resources for calculation.

At present, another disadvantage is the large time consumption to calculate rankings of feature pairs. Calculating the 2F-MRMR and the 2F-MCRMR takes about 955 s (15.9 min) and 2342 s (39.0 min). With by far larger amounts of data necessary for accurate estimation of conditional probabilities, this time will rise accordingly.

In the future, however, both enormous amounts of data as well as increased computing performance may lead to practicability and tests for reasonable results may be performed.

4.5 Relevance of Historical Measurement Values

Hitherto, this chapter introduced and investigated methods to determine the relevance of situation features, especially measurement quantities from vehicle endurance runs. Simply stated, these relevance calculations based on mutual information compare all quantity values at a certain point in time with an application function (or class) value at the same point of time. As shown, this may considerably support development and increase quality of driver assistance functions.

However, the above introduced methods are not able to consider other time points in the time series of the features relatively to each currently regarded time point. But, sometimes it may be of importance to know the relevance of historical data, i.e. looking back several seconds in time.

[Fraser and Swinney, 1986], for example, investigated mutual information within a measurement signal to determine time delays that have low mutual information with current values for phase-portrait reconstruction. It is especially intended to be used on signals containing periodic parts.

Concerning measurement data investigated in this thesis, the plain relevance of current values with respect to the class, for example, does not explain the importance of derivatives, moving averages or integrated values and time offsets in cause and effect of a signal.

For example, an objects' acceleration could be relevant to a target function and only the objects velocity may be provided. Then, because the derivative of the velocity is the acceleration, very near historical values should have high relevance as well. They are needed to calculate a derivative.

This section provides mutual information calculations based on shifted time series values of features with respect to a target function. Additionally, interpretation potential to further support function development and according issues, which have to be considered, are discussed. This discussion is illustrated by simulation results on measurement data from vehicle endurance runs.

Subsection 4.5.1 first points out some data preparation issues that are important prior to historical relevance calculation. Subsection 4.5.2 provides several measures to investigate historical relevance and subsection 4.5.2.3 discusses simulation results with these measures.

The basic idea of this approach is to use the mutual information as introduced in section 3.1 not on features among each other but on single features themselves. This means taking into account historical values and present values of the same feature. The concept is then extended to considering the feature and its historical values together with a target function (or class) feature, similarly as used in the feature rankings described earlier in this chapter.

With the approach, series of mutual information values are created over the time that is "looked back" at the feature, forming a time series. As it turns out, there may be characteristic curves for specific types of relationships within the feature and between the feature and a class, leading to assumptions on important functional relationships such as derivatives, integrals, offsets etc. These characteristic curves will be created for a number of ideal relationships. They then facilitate conclusions and assumptions on potentially existing relationships between the feature and its history towards the class.

If such relationships are discovered using this approach and seem to be useful to the user, new features built of these relationships may be generated, providing more or enhanced information towards the class. The approach furthermore offers ways to interpret characteristics of features concerning their history and recurring "behavior" over time that is statistically relevant (such as characteristic durations, time offsets, gradients etc.). During development of a system or function, it may often be useful to compare features (or signals) concerning their quality or time lag, identifying features with historical relevance to concentrate on or to find causes for effects that occur at a specific point in time prior to an event.

An alternative to identifying characteristic and possibly beneficial relationships with this approach, feature generation methods may serve to create such relationships. It may be a valid approach to find new useful features. However, a brute force method needs to be applied for feature generation to exhaustively find the right features. As an example, feature generation for a time offset feature demands the creation of all possible offsets (using a small time interval between considered offsets). Each created offset feature then needs to be tested for usefulness with mutual information (to avoid implementation of all created features in the developed system or machine learning algorithm which should in general be even less efficient). This furthermore needs to be done with all types of features taken into consideration to be generated from each original feature. The approach presented here, compared to feature generation, supports finding useful features rather directly.

In contrast, feature generation will demand tremendous time and resource complexity to find the truly useful features. Yet particularly, it does not allow for interpretation and to find effects not addressable with a predefined set of features to be generated.

The artificially generated curves for specific types of functional relationships investigated in this chapter and the interpretations on artificial and real world measurement signals provided here are a starting point and only exemplary. Further types of relationships and many further – often application-specific – interpretations are possible and may be subject of further research.

4.5.1 Preparation of Feature Time Series

Historical relevance can only be considered for time points or data points, respectively, that actually have a history. For long scenes, this is the case for most data points. However, at the beginning of the scene, there are no valid historical values for as long as the time that is to be looked back in history. For example, if the relevance of historical values laying back two seconds is to be considered, the first two seconds of each scene do not have valid historical data, because the necessary time shift is not possible. This particularly comes into effect for an increasing number of non-continuous scenes.

For the measurement data from endurance runs underlying simulations in section 4.3 (collision warning data set), there are many of these non-continuous scenes. This is for two reasons. On one hand, to gather as many valid measurement data as possible to prevent data loss by possible system errors, scenes of a length of a few minutes are saved consecutively. Between each two scenes there is a time delay causing non-continuity. On the other hand, these scenes may contain multiple objects as explained above. For each object, the part of the scene where the object occurs is repeated to gather its own data within the same features. This additional effect to the discontinuous scenes further chops the scenes. **Fig. 19** shows the number of remaining valid historical data points and valid historical object data points over the history time for the measurement data from endurance runs (section 4.3). Over time, remaining data points decrease quite significantly and object data points are by far less and relatively decrease even faster.

The issue of invalid historical data at the beginning of scenes is omnipresent with the investigation of historical relevance. To reduce related effects, it should be aimed for long, continuous scenes and data should be prepared regarding this aspect. Unfortunately, this cannot be influenced for object scenes, because they are solely depending on each object itself.

For notation, C_t determines the class or target function at each *currently* regarded time point t, F_t a feature at the same time point and F_{t-T} the same feature at the time point $t - T$ (historical feature data). Hence, if e.g. $T = 5\ s$, the feature contains the historical values of 5 s before each *currently* regarded time point. Concerning data preparation, there are two ways to handle the issue of invalid historical data at the beginning of scenes:

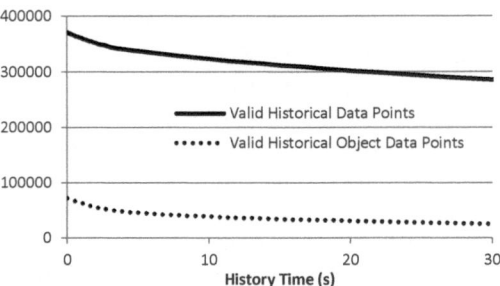

Fig. 19: Number of remaining valid data points over selected history time. The solid line shows the overall number, the dotted line shows the number of remaining data points of object scenes of the collision warning data set.

1. For the first T seconds of each scene, all history features are set to a *certain value* c_{invalid} that indicates invalid historical data:

    Let $f(\theta)$ be the values of the original feature F at time points θ, with scenes Θ_i having starting points θ_i and $\Theta = \{\Theta_i\}$ as the series of scenes, so that

    $$F = \{f(\theta): \theta \in \Theta\} . \tag{39}$$

    Then $f_t(\theta)$ is the value of the feature F_t at time point θ for history calculations at the *currently* regarded time point and in this case (because no data is actually removed)

    $$F_t = F = \{f_t(\theta) = f(\theta): \theta \in \Theta\} . \tag{40}$$

    The same applies for the class C_t with

    $$C_t = C = \{c_t(\theta) = c(\theta): \theta \in \Theta\} . \tag{41}$$

    $f_{t-T}(\theta)$ are the values of F_{t-T} with

    $$F_{t-T} = \left\{ f_{t-T}(\theta): \theta \in \Theta, \quad \begin{cases} f_{t-T}(\theta) = f(\theta - T); \ \theta \geq \theta_i + T, \ \theta \in \Theta_i \\ f_{t-T}(\theta) = c_{\text{invalid}}; \ \theta < \theta_i + T, \ \theta \in \Theta_i \end{cases} \right\} . \tag{42}$$

2. The first T seconds of each scene are *removed* from the data that is regarded for historical relevance:

    All feature series have to be of the same length for mutual information calculations and, hence, time points with invalid history have to be removed from both currently regarded features, history features and the class. The currently regarded feature as well as the class (replace F by C and f by c) are then calculated with

    $$F_t = \{f_t(\theta) = f(\theta): \theta \geq \theta_i + T, \ \theta \in \Theta_i\} \tag{43}$$

    and the values of the historical feature with

    $$F_{t-T} = \{f_{t-T}(\theta) = f(\theta - T): \theta \geq \theta_i + T, \ \theta \in \Theta_i\} . \tag{44}$$

Alternatively applying a fixed time offset over all data or setting the first value of a scene as historical data for the first T seconds of each scene both lead to wrong results. If a plain time shift is performed on measurement data, data from time points belonging to the previous scene would erroneously be considered and mutual information values would be biased. Using the first scene value for each feature at the beginning of a scene, when older values are not yet available, means, younger values of history than the actually desired historical time are considered and would distort the historical relevance as well, especially for short scenes.

Setting invalid historical data to a certain, predefined value or removing invalid data both result in losing the first T seconds on historical mutual information calculations to determine relevance of feature history. Characteristics of both preparation alternatives have to be considered during interpretation

Removing data may raise or decrease entropy and mutual information of a feature or the class, whereas setting an invalid data constant must decrease entropy. In both cases effects are unpredictable. That is, because short scenes are deleted more and more with rising historical time. Functional relationships longer in time after scene beginnings may be different. This effect decreases with growing scenes towards endless scene length, because then, less data points are deleted or invalid relatively.

For history MI values that are not normalized with other MI values (introduced in the next section), the bias may become more obvious when using a predefined invalid history value c_{invalid} at beginnings of scenes, because entropies must decrease and MI values have to follow, when history is not of high relevance. Thus, unnormalized calculations should be considered together with normalized calculations.

In general, results for historical relevance should be treated with care for higher historical times and always considered together with average lengths of scenes and remaining valid data points.

The decision, on what method for data preparation to apply, should also depend on the system underlying measurement data and the target system. When history is to be used in online system operation, an invalid history constant c_{invalid} has to be used for features at scene beginnings. This case may occur, when history has to always be provided or scenes may start over and over (e.g. object scenes). When, in contrast, the system may wait and start as soon as historical data is completely available, invalid historical time points may be removed for relevance investigation.

The following subsection will introduce useful mutual information calculations before exemplary discussing relevance of feature history based on measurement data.

4.5.2 Determination of Historical Relevance with Mutual Information

This thesis proposes mainly three types of historical mutual information, with historical feature values F_{t-T}, *currently* regarded feature values F_t and a target function C_t, useful to be calculated to indicate the relevance of historical data:

$$I(F_{t-T}; F_t), \ I(F_{t-T}; C_t), \ I(F_{t-T}, F_t; C_t) \ . \tag{45}$$

For the definition and preparation of the variables F_{t-T}, F_t and C_t, please refer to subsection 4.5.1.

These measures in (45) go beyond the plain intra-signal time series investigation of [Fraser and Swinney, 1986].

The three types will be explained and interpreted in detail in the next subsections. It is of particular interest to be able to compare the importance of historical data between different features. For comparability, the latter two types of mutual information $I(F_{t-T}; C_t)$, $I(F_{t-T}, F_t; C_t)$ are therefore normalized using the mutual information of the *current* feature with the *current* target function $I(F_t; C_t)$ at time t. The first type $I(F_{t-T}; F_t)$ is normalized with the entropy of the *current* feature $H(F_t)$.

To illustrate and discuss effects and curves of all three types of normalized mutual information, simulations on artificially generated functional relationships where carried out. They are briefly introduced before the types of normalized history MI are explained.

4.5.2.1 Characteristic Patterns of Artificially Generated Functions

Effects of historical relevance are outstanding on four types of signals and some variations: derivatives, offsets, moving averages and periodic signals. As a basis, an approximately Gaussian-distributed real-world measurement signal is picked as input feature with values $s(\theta)$, such that in this subsection $f(\theta) := s(\theta)$ is used. In this thesis this signal is preferred over an artificial, random-value-based signal, because mutual information with a pure, equally distributed random signal (white noise) will always be zero. Even, if a time series with values depending on previous values is built, which uses randomized values for the signal change, such effects cannot be ruled out and the result may be biased[31]. For the periodic signals, trigonometric functions are used.

As target classes or functions for mutual information calculations, several functions $c(\theta)$ are calculated. They are based on the real world measurement signal $s(\theta)$:

- Two derivatives c_{Dt15} and c_{Dt3}, curves shown in **Fig. 20**, with

$$c_{Dt\alpha}(\theta) = \frac{s(\theta) - s(\theta - \alpha)}{\alpha} \ . \tag{46}$$

- Two rolling averages (approximate integrals) c_{Avg50} and c_{Avg5}, curves shown in **Fig. 21**, with

$$c_{Avg\alpha}(\theta) = \frac{1}{\alpha} \sum_{i=0}^{\alpha} s(\theta - i) \ . \tag{47}$$

- Two time offsets c_{Off50} and c_{Off5}, curves shown in **Fig. 22**, with

---

[31] This will particularly come to effect for historical mutual information calculations, because the relationship to previous values is investigated. This relationship to previous values is then, by definition, randomized and thereby close to 0.

$$c_{\text{Off}\alpha}(\theta) = s(\theta - \alpha) .$$ (48)

Because periodic signals have interesting characteristics, the following periodic features and target functions are investigated as well:

- A pure periodic sine signal $f_{\text{sin}100}$ and a cosine target function $c_{\text{cos}100}$ and a noisy sine signal $f_{\text{sin}100,\text{noise}}$ and a noisy cosine target function $c_{\text{cos}100,\text{noise}}$, curves shown in **Fig. 23**, with

$$
\begin{aligned}
f_{\text{sin}100}(\theta) &= \sin\left(\theta \cdot \frac{2\pi}{100}\right) , \\
c_{\text{cos}100}(\theta) &= \cos\left(\theta \cdot \frac{2\pi}{100}\right) , \\
f_{\text{sin}80,\text{noise}}(\theta) &= 0.2 \, \text{rdm}() + \sin\left(\theta \cdot \frac{2\pi}{100}\right) , \\
c_{\text{cos}80,\text{noise}}(\theta) &= 0.1 \, \text{rdm}() + \cos\left(\theta \cdot \frac{2\pi}{100}\right) .
\end{aligned}
$$ (49)

The function rdm( ) generates (pseudo-)random values between 0 and 1.

- Similar noisy, periodic signals containing decreasing historical relevance, curves shown in **Fig. 24**, with

$$
\begin{aligned}
f_{\text{sin}40,\text{RampFast}}(\theta) &= 0.050 \, \text{rdm}() + \sin\left(\theta \cdot \frac{2\pi}{40 + \theta \cdot 2.5 \cdot 10^{-5}}\right) , \\
c_{\text{cos}40,\text{RampFast}}(\theta) &= 0.025 \, \text{rdm}() + \cos\left(\theta \cdot \frac{2\pi}{40 + \theta \cdot 2.5 \cdot 10^{-5}}\right) , \\
f_{\text{sin}40,\text{RampSlow}}(\theta) &= 0.025 \, \text{rdm}() + \sin\left(\theta \cdot \frac{2\pi}{40 + \theta \cdot 6.5 \cdot 10^{-6}}\right) , \\
c_{\text{cos}40,\text{RampSlow}}(\theta) &= 0.010 \, \text{rdm}() + \cos\left(\theta \cdot \frac{2\pi}{40 + \theta \cdot 6.5 \cdot 10^{-6}}\right) .
\end{aligned}
$$ (50)

Characteristics of these artificially generated curves in **Fig. 20** to **Fig. 24** are discussed together with the types of normalized historical mutual information in the following subsections. **Fig. 27** and **Fig. 28**, by contrast, show simulations on the collision warning data set from section 4.3. They are discussed later in section 4.5.2.3.

4.5.2.2 Normalized Historical Relevance Measures

Normalized Relevance of the Historical Feature Value F_{t-T} at Time $t - T$ with the Current Feature Value F_t

This measure $I_{\text{rel,Consistency}}$ describes how much information about the historical feature values is still contained in the current feature values and vice versa. In other words, it tells how predictable future values are or how dynamically the feature changes over time.

It is defined with[32]

$$I_{\text{rel,Consistency}}(T) = \frac{I(F_{t-T}; F_t)}{H(F_t)} .$$ (51)

Different curves for $I_{\text{rel,Consistency}}$ are illustrated in **Fig. 20** to **Fig. 28** (blue solid or dotted lines).

---

[32] For readability, $I_{\text{rel,Consistency}}$ is labeled MI(Curr;History) in figures.

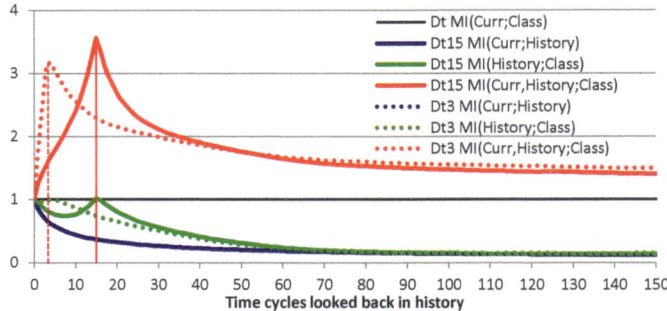

Fig. 20: Simulated curves of historical relevance for artificially generated derivatives calculated with differences of 15 (solid line) and 3 (dotted line) time cycles (see formula (46)).

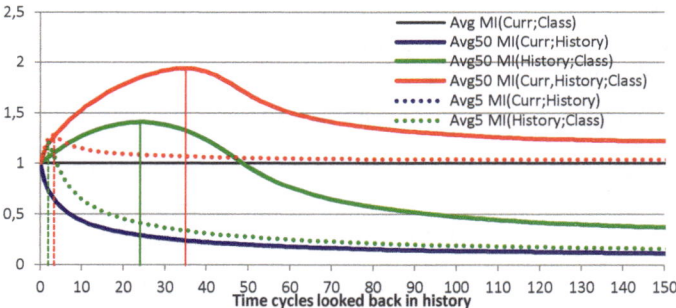

Fig. 21: Simulated curves of historical relevance for artificially generated rolling averages calculated over 50 (solid line) and 5 (dotted line) time cycles (see formula (47)).

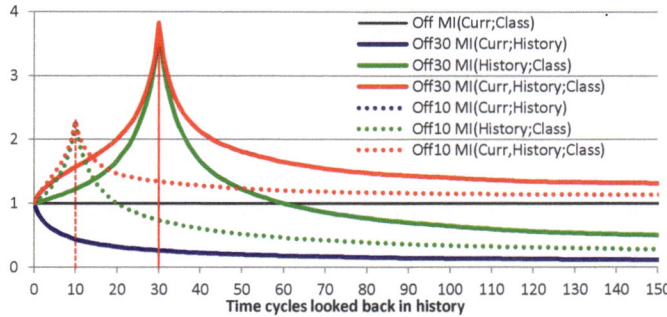

Fig. 22: Simulated curves of historical relevance for artificially generated time offsets of the feature itself calculated with 30 (solid line) and 10 (dotted line) time cycles (see formula (48)).

For a pure and periodic signal, $I_{\text{rel,Consistency}}(T)$ would stay high and become 1 for values T being integral multiples of the periodic length because for these time points the signal is perfectly predictable. In special cases there may be several peaks for fractions of T, e.g. if the periods are symmetric or rotationally symmetric (for generated artificial pure periodic signals with and without noise see **Fig. 23**, for artificial signals with decreasing periodic influence see **Fig. 24**). In even more special cases, $I_{\text{rel,Consistency}}(T)$ may completely remain 1, if the signal is periodic and there is no ambiguity about any future value for a given T. An example of such a signal is a saw tooth pattern, as it is used e.g. for time signals.

For a less dynamic signal, this value would slowly decrease over T, whereas for a highly dynamic signal, this value would decrease rapidly over T. In the special case of a periodic signal, the value would remain high even for signals with high dynamics (see **Fig. 23**).

Note, that in the general and theoretic case, it is always $I(F_{t-T}; F_t) \leq H(F_t)$ and thus $I_{\text{rel,Consistency}}(T) \leq 1$ for all T. This still holds, if invalid historical data is assigned with a certain value c_{invalid} indicating invalidity as in (42). This lowers the entropy of the history feature and the inequality $H(F_{t-T}) \leq H(F_t)$ remains true. Unfortunately, this cannot generally be said, if invalid historical data is removed as in (43), (44). In this case, the content, that the entropies $H(F_t)$ and $H(F_{t-T})$ are based on, does not remain the same.

Normalized Relevance of the Historical Feature Value F_{t-T} with the Current Function Value C_t

It is of even greater interest than the direct comparison of the current and the historical feature how much information the historical feature actually provides to the target function. Divided by the mutual information of the current feature with the target function, the measure $I_{\text{rel,PlainHist}}(T)$ directly illustrates the relative loss or gain of information, when plainly using the history feature compared to using the current feature. This has effect also for time delays in processing of measurement data.

The measure is defined with[33]

$$I_{\text{rel,PlainHist}}(T) = \frac{I(F_{t-T}; C_t)}{I(F_t; C_t)} \quad . \tag{52}$$

Values of $I_{\text{rel,PlainHist}}(T)$ remaining high over T generally indicates high importance of historical data to the target function and potentially lower signal dynamics. A rapidly decreasing value, in contrast, means low importance and potentially higher signal dynamics (compare green solid and dotted curves in **Fig. 20** and **Fig. 21**).

A special and very important case is $I_{\text{rel,PlainHist}}(T) > 1$ for some T (green curves in **Fig. 21**, **Fig. 22** and **Fig. 23**). This implies historical values are actually more relevant to the target function than current feature values. A single (local) peak generally points towards a time offset of cause and effect in the relationship of the feature and the target function (**Fig. 22** and **Fig. 23**). Values greater 1 over a wider range of T indicate rather general importance of history (**Fig. 21**). Such a signal does have predictive character with respect to the target function.

---

[33] For readability, $I_{\text{rel,PlainHist}}(T)$ is labeled MI(History;Class) in figures.

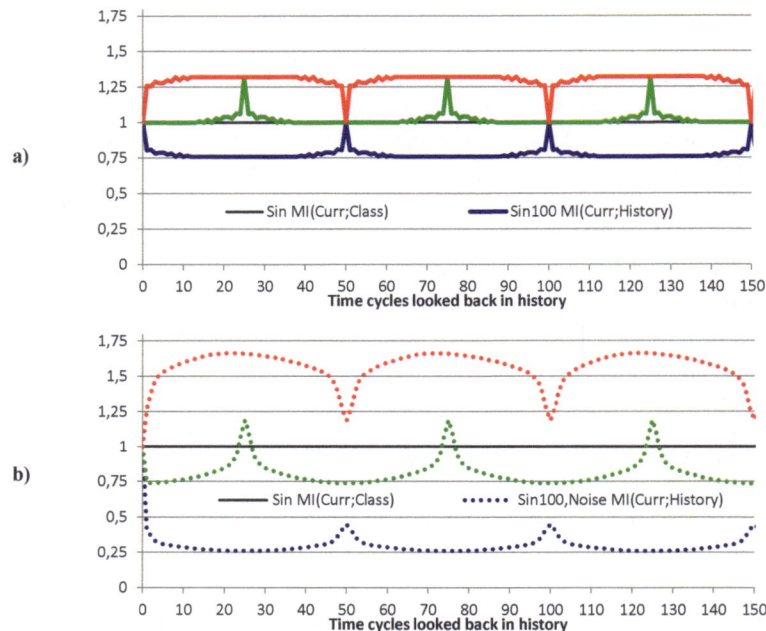

Fig. 23: Simulated curves of historical relevance for artificial pure periodic features **a)** without (solid curves) and **b)** with (dotted curves) noise and periodic lengths of 100 time cycles (as in (49)).

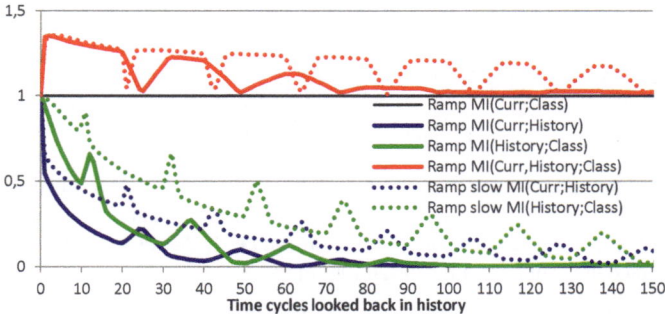

Fig. 24: Simulated curves of historical relevance for artificial features with fast and slowly decreasing periodic influence and periods of 40 time cycles, respectively (as in (50)).

For example, a properly used direction indicator is a predictive signal for a future lane change. Similarly, a yellow flashing light may be a warning of a potentially critical situation ahead. Thus, there will be a strong relationship of current values of these features with future values of the function (prediction), but less relationship of future values of these features with their future function values. If these future times are shifted to be the current time points – this makes the future function value the current function value – the formerly current feature values (e.g indicator signal) become historical values. These historical values then have a stronger relationship with the current function value than the current feature values. In this way, the mutual information using these historical feature values is higher than the mutual information using current feature values.

Furthermore, signals with several peaks (as in **Fig. 23** and **Fig. 24**) may be observed. Several reasons may cause this type of signal sequence. In general, several peaks point out several offsets in time with respect to predictive relevance.

For example, assuming that vehicles drive at typical speeds, distributed around typical speed limits, several peaks would occur at time offsets resulting from these typical speeds together with distances ahead of locations relevant to a target function. Let us think of a sensor that recognizes critical objects at distances of 50 m[34]. Vehicles may drive at typical exemplary speeds of 30, 50 and 70 km/h. Then, the sensor will typically recognize critical objects 6.0, 3.6 and 2.6 s before passing, or potentially hitting, the critical object. Historical relevance calculation in this example should show peaks or at least rises in mutual information at historical times of 6.0, 3.6 and 2.6 s and below.

A special case of an offset is a periodic signal. A signal of this type will have several local peaks at integral multiples of the periodic length or fractions of the periodic length similar to the explanation above with $I_{rel,Consistency}(T)$ having local peaks for periodic signals. A pure periodic signal will have peaks at each integral multiple of the periodic length (**Fig. 23**).

Normalized Joint Relevance of both the Historical Feature Value F_{t-T} *and* the Current Feature Value F_t with the Current Function Value C_t

The case that historical data has actually more relevance to a target function than current data is assumed to be rather rare. Generally, data should be as up to date as possible, which is the case, when values of $I_{rel,PlainHist}(T)$ are continually decreasing with $I_{rel,PlainHist}(T) < 1$ for all $T > 0$ (illustrated by **Fig. 24** and the dotted green curve for feature 4 in **Fig. 27** and **Fig. 28**).

However, the relevance of historical data is usually not completely 0 or entirely redundant to current data and the use of historical data in addition to current data will increase the information with respect to a target function.

Hence, the joint mutual information $I(F_{t-T}, F_t; C_t)$ of both the feature containing current data and the feature containing historical data with respect to a target function and the normalized relative measure $I_{rel,JointHistory}(T)$ are of interest.

---

[34] This is only a simple arithmetic example for illustration.

The measure is defined with[35]

$$I_{\text{rel,JointHistory}}(T) = \frac{I(F_{t-T}, F_t; C_t)}{I(F_t; C_t)} \quad . \tag{53}$$

It is always greater than or equal to the mutual information between the current data feature and the target function: $I(F_{t-T}, F_t; C_t) \geq I(F_t; C_t)$. Subsequently, it is $\forall T \geq 0$: $I_{\text{rel,JointHistory}}(T) \geq 1$, which follows from (9) and (10), respectively.

This joint mutual information, as included in (53), and the according relative quantity provide a measure for the relevance of both current data and historical data from a certain, relative point of time in the past.

For $T = 0$, the joint mutual information is equal to the mutual information with plainly the current feature. For an increasing T, the joined historical feature will more and more differ from the current feature and add new information for the target function (taken from the past). Eventually, after reaching a peak (or several local peaks) the joint mutual information will decrease again with the historical data laying too far in the past to deliver any relevant information for the target function. Because the mutual information including the current feature still remains, in the general case the joint mutual information will approach this value again.

When interpreting the joint mutual information of both current and historical data, as in (53), functional relations requiring both current and historical data can be discovered. This is an advantage over plainly interpreting historical data, as in (52). There are many possible functional relations using data with a time offset, such as derivatives, integrals, moving averages, relative changes etc. This, in particular, cannot be achieved by considering plain feature relevance as hitherto investigated in this chapter and introduced in section 3.1.

A derivative is an operation on contiguous sequence values very close in time. Hence, a high relevance of a features' derivative is indicated by a high peak around $T \approx 0$, but not equal to 0, especially within one or only a few time intervals of sequence data (compare **Fig. 20**). If the peak lies around several but still few time intervals, the derivative may be required to be smoothed or filtered or a moving average gradient may be of interest (compare also **Fig. 21**). Note, that for derivatives, the normalized plain history MI $I_{\text{rel,PlainHist}}(T)$ stays high over the interval, which the derivative is built over, and reaches 1 again at the second time point. Both time points (used in (46)) equally contribute to the target function.

The dotted curve in **Fig. 27** (feature 4) represents the rather general case when the relative joint mutual information is not too high above 1, indicating (relatively) little importance of historical values[36].

Some features may show a particularly high relative information gain with a plateau over a wider range of historical time T (compare the artificial examples). For these features, functional relationships with several historical values or historical values with a bigger distance in

---

[35] For readability, $I_{\text{rel,JointHistory}}(T)$ is labeled MI(Curr,History;Class) in figures.

[36] Be aware, that this concerns relative mutual information discussion. The absolute joint mutual information is recommended to always be considered in combination. The relative value may be low, but the absolute value high in comparison to other features, if $I(F_t; C_t)$ is very high for this feature.

time are assumed to be of interest. Such functional relationships may be integrated features or rolling average features or the gradient of rolling average features. It is recommended to subsequently generate features of this type and determine their relevance additionally.

The other curves in **Fig. 27** and **Fig. 28** (features 1 to 3) are examples for features, where history *is* of interest. Several peaks are suggestive of important historical time values, potential periodicity or further characteristics in the relationship between features and target functions as explained above.

4.5.2.3 Historical Relevance of Artificial Signals w. r. t. Discrete Classes

Fig. 20 to **Fig. 24** showed typical curves for artificially generated functions, which are formally defined by (46) to (50). These curves are aimed to be used for the interpretation of real-world measurement signals. If the characteristics of these curves are recognized, a feature with the according functional relationship to the original feature may be generated.

However, target classes, that are used for historical relevance calculation, are not continuous, like the artificially generated functions, and do not have this kind of exact functional relationship.

To still be able to transfer typical characteristics to real-world signals and mutual information w. r. t. a discrete class, simulations with the artificially generated functions were rerun. But this time, each function was scaled and rounded to form only four discrete classes. This is aimed to approximate the case with a strong relationship of the kind of the artificially generated function on a real-world class with only few discrete labels.

The resulting curves, using only four target discretizations with a maintained strong relationship of the according type of artificially generated function, are shown in **Fig. 25**, **Fig. 26**. Obviously, the according curves do not change significantly. It can be assumed that similar characteristics would be recognizable in historical relevance investigations on real-world measurement signals. This type of discussion is shown in the next section at the example of four signals that historical relevance is calculated with.

4.5.3 Simulation Results on Historical Relevance of Features

Four feature examples taken from measurement data hitherto investigated with respect to relevance rankings (collision warning data set from section 4.3) will be subsequently shown and discussed regarding their historical information. To support unprejudiced interpretation and avoid presumptions, signals are discussed anonymously.

Fig. 27 and **Fig. 28** show the previously introduced measures $I_{\text{rel,Consistency}}(T)$ from (51) in blue, $I_{\text{rel,PlainHist}}(T)$ from (52) in green and $I_{\text{rel,JointHistory}}(T)$ from (53) in red for four features plotted over the historical time offset T (solid or dotted for distinction). These features deliver some nicely interpretable results to be exemplary discussed and compared. In this case, history features were created by removing invalid historical data points as in (43), (44).

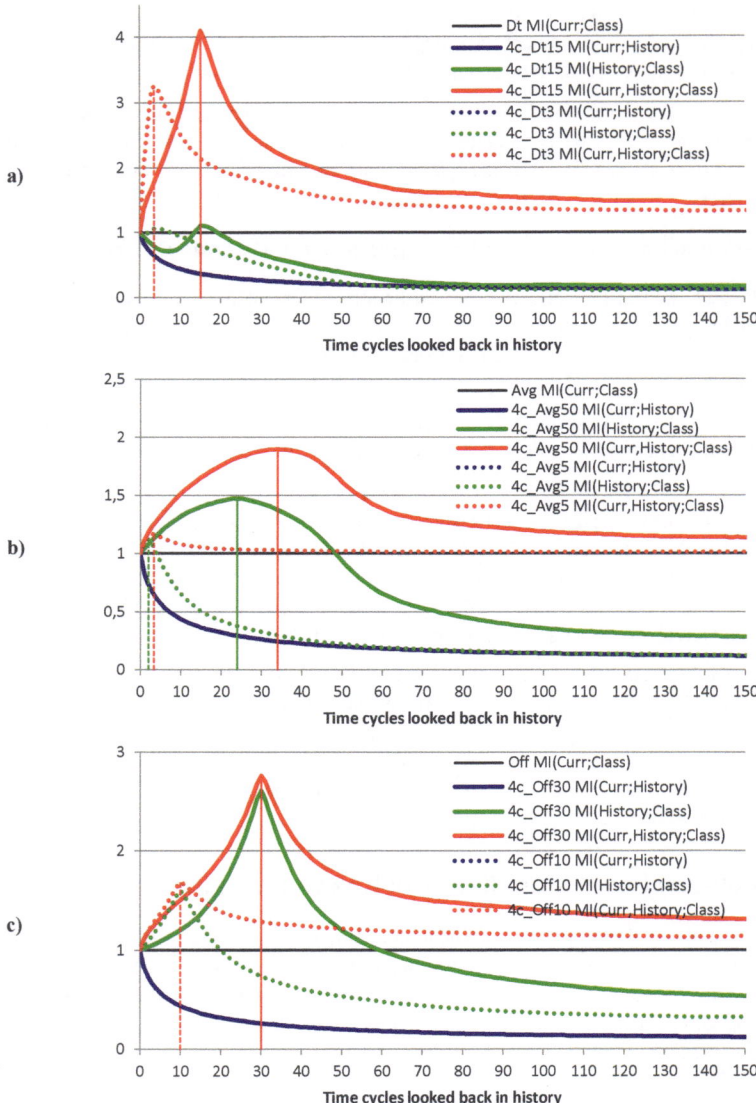

Fig. 25: Curves of historical relevance of class functions (as in **a)** (46), **b)** (47) and **c)** (48)) with each class function rounded down to only four discrete classes. These plots are meant to be compared directly with **a) Fig. 20**, **b) Fig. 21** and **c) Fig. 22**.

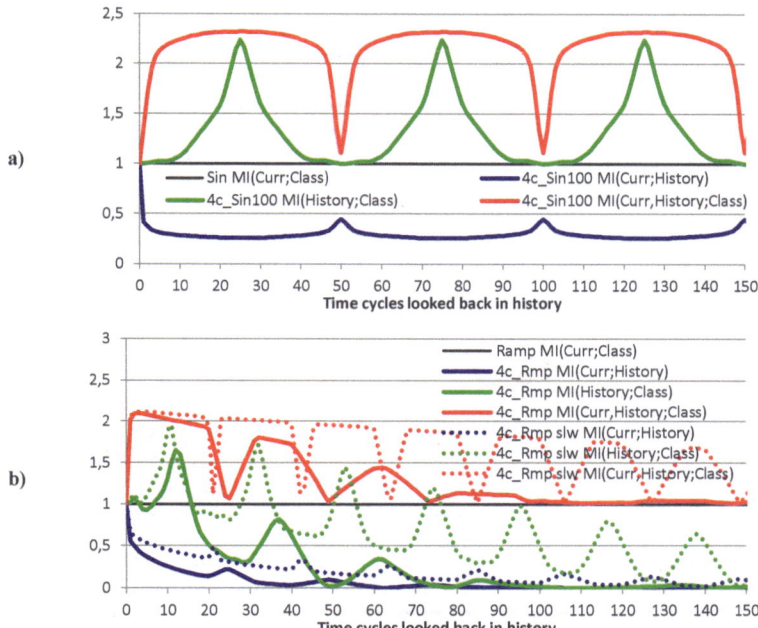

Fig. 26: Curves of historical relevance of class functions (as in **a)** (49) and **b)** (50)) with each class function rounded down to only four discrete classes. These plots are meant to be compared directly with **a) Fig. 23** and **b) Fig. 24**.

In **Fig. 27**, the historical relevance measures $I_{rel,PlainHist}(T)$ and $I_{rel,JointHistory}(T)$ are normalized with $I(F_t; C_t)$. Removing invalid historical data is hence considered in normalization, so that the inequality $I(F_{t-T}, F_t; C_t) \geq I(F_t; C_t)$ holds (red curves stay above 1).

In contrast, these measures are normalized with $I(F; C)$ in **Fig. 28**. This considers the original features F and the original class C. The inequality does not hold. The plots in **Fig. 28** hence also show how the mutual information $I(F_t; C_t)$ behaves over T, when invalid historical data is removed, that is, only scenes longer than T are considered. This is plotted by the black curves in **Fig. 28**.

There is a difference in meaning of the curves plotted in both figures. **Fig. 27** illustrates the relevance of history *in* the remaining scenes (scenes longer than T), whereas the overall relevance of history compared to the original feature relevance is illustrated in **Fig. 28**.

For example at $T = 10\ s$, a joint historical relevance greater than 1 in **Fig. 27** means considering a history feature at $T = 10\ s$ improves classification[37] in all scenes longer than 10 s.

---

[37] provided a classifier is learned with this historical and the original feature

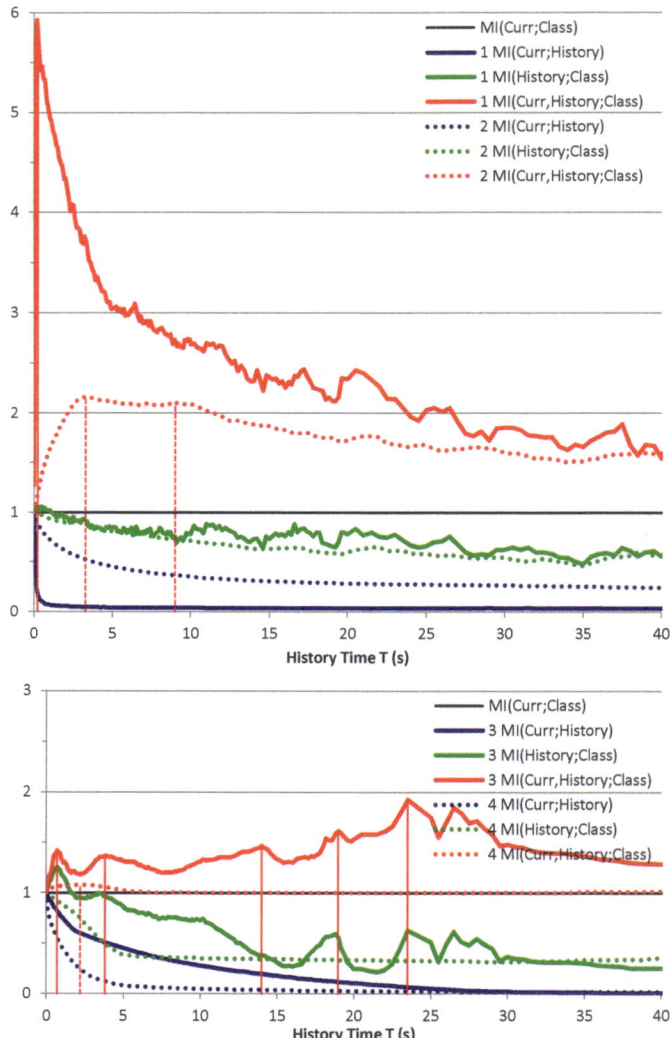

Fig. 27: *Relative* historical mutual information relevance for several real world measurement features from vehicle endurance runs (equal axis scales).
The historical relevance measures are normalized with $I(F_t; C_t)$, the mutual information regarding only valid historical data, so that asymptotic behavior towards 0 or 1, respectively, is ensured, but only the gain w. r. t. the remaining valid data is considered.

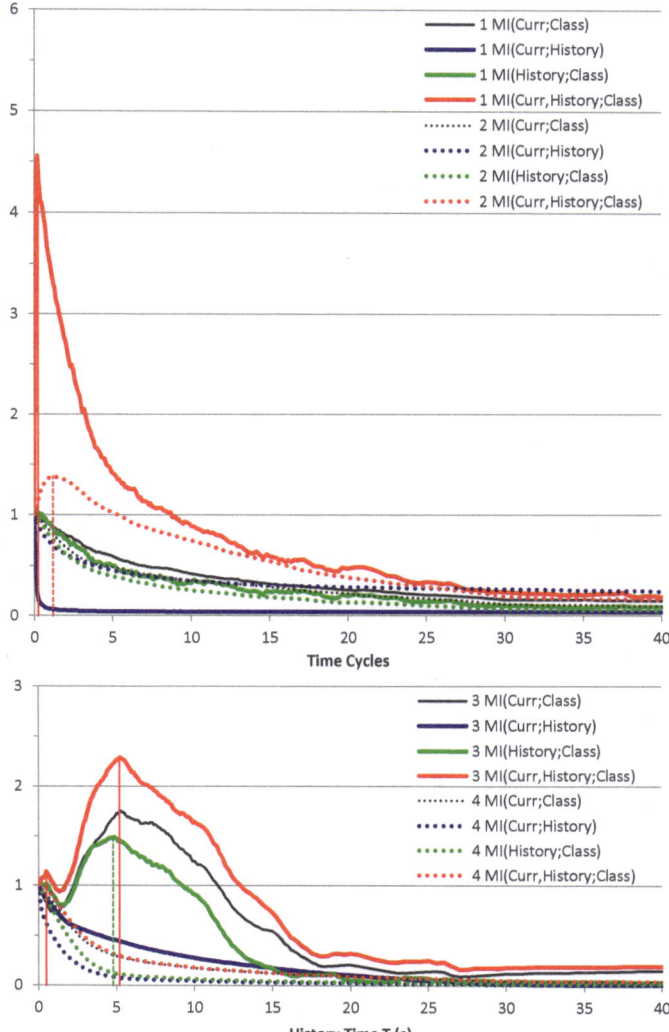

Fig. 28: *Relative* historical mutual information relevance for several real world measurement features from vehicle endurance runs (equal axis scales).
The historical relevance measures are normalized with $I(F; C)$, the mutual information regarding the original data, so that asymptotic is *not* ensured, but the overall gain is considered.

However, a joint historical relevance greater than 1 in **Fig. 28** would even improve overall classification of all scenes (although there is no effect on shorter scenes).

Note, that these curves are examples of a data set based on real world measurements in the automotive domain. They may vary significantly for other types of features and their historical relevance results.

4.5.3.1 Application-Independent Interpretation

In comparison to features 1 to 3, feature 4 shows an example where history has hardly any relevance. Plain historical relevance $I_{\text{rel,PlainHist}}(T)$ decreases fast and stays on a low level and $I_{\text{rel,JointHistory}}(T)$ keeps close to 1 in **Fig. 27** or below 1 in **Fig. 28**.

Fig. 28 shows, that above $T = 20\ s$, $I(F_t; C_t)$ is quite low and does not change much more for all features. Historical relevance measures, especially those normalized with $I(F_t; C_t)$ in **Fig. 27** should therefore only be interpreted with care for $T > 20\ s$, especially for feature 3.

Historical Consistency

Regarding the consistency measure $I_{\text{rel,Consistency}}(T)$, the features differ significantly. For example, this value rapidly decreases for feature 1, denoting high dynamics of the signal, whereas it decreases more than 50 times slower for feature 3. This implies much lower dynamics and seldom changes of the signal state or value, respectively. As there are no peaks or higher levels of this value for all four features, they seem to have no periodicity.

Plain Historical Relevance

When having a look at the sole relevance of historical data for the target function, all features 1 to 3 have slowly decreasing and relatively high values of $I_{\text{rel,PlainHist}}(T)$. So, independently from the signal dynamics, historical values of the features themselves are relatively important, particularly for features 1 and 3. Especially, for all 3 features, there is a region of up to $T = 10\ s$ with still high historical relevance, considering both **Fig. 27** and **Fig. 28**. This region may be interesting for average building and signal filtering.

Feature 3 comes along with additional peculiarities exposing a very high predictive capability. Note, however, that the absolute mutual information $I(F_t; C_t)$ or $I(F_{t-T}, F_t; C_t)$, respectively, has to be taken into account to interpret the complete importance of history.

For this feature 3, it is exceptionally remarkable that $I_{\text{rel,PlainHist}}(T) \gtrsim 1$ up to $T = 3.0\ s$ with a peak of about 1.25 at $T = 0.7\ s$ in **Fig. 27**. More interestingly, the overall effect normalized with the original mutual information in **Fig. 28** is even higher with a peak of 1.49 at $T = 4.8\ s$ and $I_{\text{rel,PlainHist}}(T) \gtrsim 1$ up to roughly $T = 10\ s$. This means, history for this feature is actually more important than the current value with the best predictability at around $T = 5\ s$ and some short term predictability at about $T = 0.7\ s$. Thus, the current state of the feature provides most information about the future situation 5 s ahead[38]. Analogy with curves in **Fig. 25 b)** and **c)** is remarkable. This indicates, derived features with a time offset or filtering such as average building are of interest.

At $T = 3.0\ s$ and to some extent at around $T = 14\ s$, $T = 19\ s$ and $T = 23\ s$ further peaks appear in joint and plain historical relevance in **Fig. 28**. Noticeably, this feature has overall high relevance of historical data up to at least 30 seconds with several peaks. According features might be generated. However, as explained in the beginning of this subsection, such characteristics should be interpreted with care due to the low feature-class-MI $I(F_t; C_t)$ that is used for normalization. Mind the effect of biased calculations of historical relevance with increased historical time. Nonetheless, the peaks could be considered and investigated. They may as well point out interesting effects laying back in history.

Joint Historical Relevance

Regarding the curves of $I_{rel,JointHistory}(T)$ of feature 1, the similarity to the dotted curve in **Fig. 25 a)** of the rounded artificially generated derivative function is remarkable. Plain historical relevance is close to 1 at around $T = 0.2\ s$ and joint historical relevance shows a very high peak at the same point of historical time. The derivative of feature 1 is thereby assumed to have a high benefit and should be taken in consideration.

Feature 2 has high values of $I_{rel,JointHistory}(T) \approx 2.1 \gg 1$ at around $T = 3.3\ s$ in **Fig. 27**, which stay high over a wide range. In **Fig. 28**, the feature has $I_{rel,JointHistory}(T) > 1$ up to roughly $T = 5\ s$. This may be used either for machine learning, using both the current and some history features within this time range or for feature generation considering relevant historical time points. Note, that for this feature, both the current and the historical value always have to be considered. Plainly considering the history itself does not have a higher mutual information than the current data, in contrast to feature 3. This high importance of short term history suggests benefit using moving averages, their derivatives or similar derived features.

Feature 3, in contrast to feature 2, has several local peaks. **Fig. 27** shows a steadily increasing value of $I_{rel,JointHistory}(T)$ up to $T = 24.4\ s$. The first peak corresponding to the maximum of the plain historical relevance $I_{rel,JointHistory}(T) \approx 1.4$ is at $T = 0.7\ s$. Two other ones can be found at $T = 4.0\ s$ and around $T = 13\ s$. These may be predictive offsets and may point out typical time intervals before certain events.

Remarkably, in **Fig. 28**, this effect is even higher with a peak of $I_{rel,JointHistory}(T) \approx 2.3$ at $T \approx 5.0\ s$. Interpretation has already been discussed for its plain historical relevance.

Filters detecting signal changes and providing signal dynamic measures may be useful within this time range.

---

[38] Note, that this is not an overall statement and does not consider all features in comparison, but historical relevance comparison concerning this specific feature.

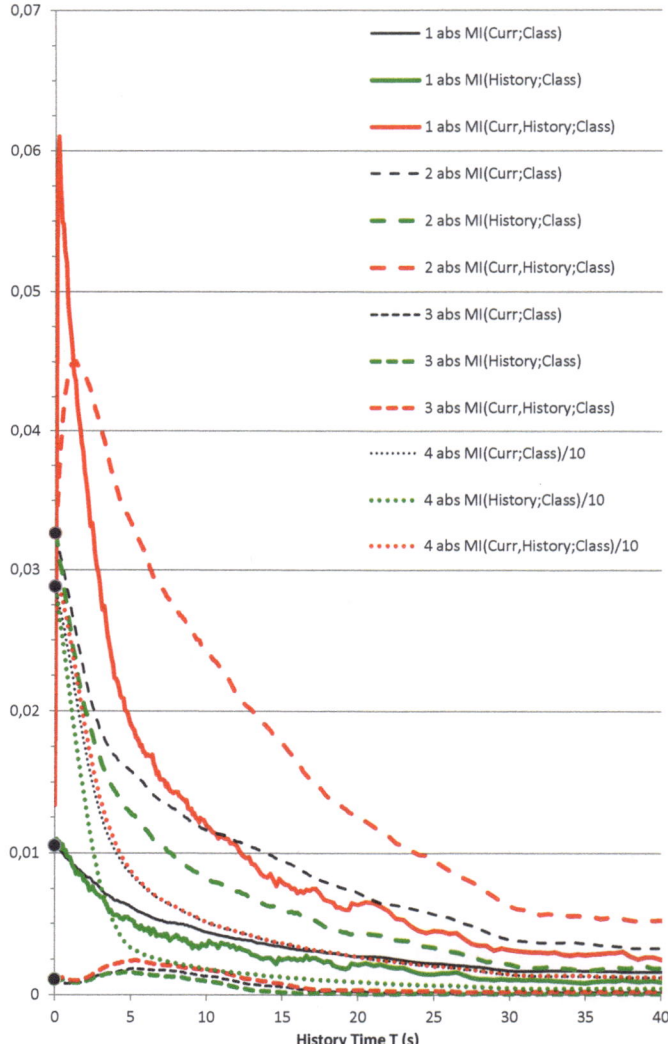

Fig. 29: *Absolute* historical mutual information relevance for several real world measurement features from vehicle endurance runs.
Curves have the shape of those in **Fig. 28** without being normalized.
(Mutual information of feature 4 divided by 10. Black circles illustrate original feature-class-MIs.)

Table 7: Names of features that were anonymized beforehand in **Fig. 27** to **Fig. 29** and their maximum relative and absolute historical MI gains.

Feature no.	Feature name	Max. rel. MI gain (see Fig. 27)	Max. rel. MI gain (see Fig. 28)	Max. abs. MI gain (see Fig. 29)
1	*GasPedal*	4.92 (0.2 *s*)	3.55 (0.2 *s*)	0.0505 (0.2 *s*)
2	*EgoVx*	1.16 (3.3 *s*)	0.38 (1.2 *s*)	0.0124 (1.2 *s*)
3	*DirIndR*	0.92 (24 *s*)	1.29 (5.2 *s*)	0.0013 (5.2 *s*)
4	*TTC*	0.08 (2.2 *s*)	0.01 (0.1 *s*)	0.0027 (0.1 *s*)

4.5.3.2 Application-Specific Interpretation

So far, the features in **Fig. 27** and **Fig. 28** were held anonymous. This made a general discussion about curve characteristics and historical relevance possible and the type of feature and according assumptions did not overshadow an objective interpretation.

The features were taken from the collision warning data set from section 4.3 (data description in subsection 4.3.2). Historical relevance was therefore determined with respect to the de-escalation maneuver class (subsection 4.3.3), so that it is possible to draw comparison to the feature relevance rankings investigated in this chapter. The concrete feature names are provided in **Table 7**.

Moreover, the maximum relative joint historical MI gain as visible in **Fig. 27** and **Fig. 28**, that is the added relative mutual information, and the maximum absolute joint historical MI gain are provided together with their offsets in time in the last two columns. The maximum absolute MI gains are visible in the plots about absolute historical mutual information in **Fig. 29**.

Some interesting feature-dependent characteristics are discussed in the remainder of this subsection.

The highest effect of historical relevance both in relative and absolute MI gain can be found for the feature *GasPedal*. It was discussed above that the derivative of this feature is assumed to be of high benefit. The collision warning data set contains a feature *GasPedalDt*, which is taken from in-vehicle signals. Therefore, this feature was not created as a result of historical relevance calculations. If, however, the derivative of *GasPedal* is of this high relevance, this should be provable by the feature *GasPedalDt*. Mutual information of both features is provided in **Table 3**. There, it is $I(GasPedal; C_t) = 0.0105$ and $I(GasPedalDt; C_t) = 0.0056$. This shows that the feature *GasPedalDt* by itself does not provide more information. However, consider that the joint historical relevance contains information of both the current *and* the historical feature. Thus, the effect of high joint historical relevance can only be seen in the joint MI of both *GasPedal* and *GasPedalDt*. This joint MI is provided in **Table 9** and amounts to $I(GasPedal; GasPedalDt; C_t) = 0.0388$. The MI gain based on the feature *GasPedal* is $I(GasPedal; GasPedalDt; C_t) - I(GasPedal; C_t) = 0.0388 - 0.0105 = 0.0283$. The joint MI is almost four times the original MI of *GasPedal* and almost seven times the original MI of *GasPedalDt*. This does not correspond to the full possible gain provided in **Table 7**, but is similarly high. Consider that it is unknown how the feature *GasPedalDt* was exactly generated. The feature may have been generated with much lower time gap, whereas the maximum in historical relevance occurs at around $T = 0.2\ s$. Overall, this high joint MI gain of both features indicates that historical relevance calculation is useful and the interpretation was correct and yields high benefit for the feature *GasPedal*.

DirIndR, the feature of the right direction indicator, contains further interesting effects. However, although there is relatively high predictive capability and relative historical relevance, this feature has hardly any mutual information with the class. This is shown by the comparison of the plots for absolute mutual information in **Fig. 29**. The plausible fact that the direction *indicator* serves to predict future situations, which it was actually invented for, is confirmed by the historical relevance plots in **Fig. 27** and **Fig. 28**. Typically, the direction indicator is used within a few seconds before an intended action. This corresponds to the high plain historical MIs in the time range up to $T = 10\ s$ with several local peaks.

Fig. 27 though contains peaks at historical times T much higher as far as up to $T = 30\ s$. It was pointed out that these peaks may be caused by biases in historical relevance calculation. But let us assume, these peaks denote real effects. Thinking of when indicators are frequently used leads to typical situations at traffic lights. During an approach to a traffic light the indicator is set and typically this state is kept until passing the traffic light. Traffic light phases, in addition, are usually quite long up to more than $30\ s$. This may be an explanation for the peaks at historical times laying this far back. They may indicate that traffic light phases of around 15, 20 and 25 s could be typical.

4.6 Summary

As described, sensors and systems in an automotive vehicle today provide and need a vast amount of signals or features, particularly advanced driver assistance systems. Determining appropriate features to be used in such a system is a task nowadays hardly tangible for the human mind or that of an automotive systems developer, respectively.

To deal with this issue and to facilitate the approaches introduced in the remaining chapters of this thesis, this chapter elaborated the application of mutual information based feature selection to provide an effective and automated way of performing this task. It could be shown to work highly efficient when being used on vast amounts of automotive measurement data. It helps to keep state space small in the subsequent function development, fed machine learning algorithms or modeled knowledge through a goal-oriented selection of features. Using modeled knowledge for driver assistance functions is subject of the next chapter.

The results can help to find suitable features and to support their target-oriented generation for the development of advanced driver assistance system functions. Moreover, tremendous benefit of these methods over the selection of features by experts of the automotive community involved in the development of driver assistance systems was shown. It was also shown with respect to one method that potential to even further improve mutual information based methods persists.

So far, methods and research with respect to mutual information feature selection was not much concerned with the history of features and their relation or contribution to the present situation. The last section in this chapter was hence dealing with the relevance of historical data and how such an approach can support automotive systems development. The introduced approach has the ability to show characteristic curves for specific types of relationships between the feature and its history leading to assumptions on important functional relationships such as derivatives, integrals, offsets etc. Having discovered such potential functional relationships a developer is able to generate new specific features or make further interpretations about the importance of a features history and possible root causes of an observed effect. Compared to exhaustive, time and resource consuming feature generation, using mutual information on historical features may sometimes be an easier way or provide an alternative to generate specific useful features or make interpretations.

Chapter 5

Knowledge-Based Traffic Situation Description

Beginning with section 5.1, this chapter describes the knowledge engineering and structure of a generic traffic situation description ontology that was motivated in section 2.2.6. Aspects relevant to a generic situation description, as pointed out in section 2.2.3, are moreover discussed on the developed ontology. The introduced ontology is designed for complex traffic situations, especially those at intersections.

Subsequently, section 5.2 presents a real-time framework, which, in section 5.3, is shown to successfully run different DAS systems, making use of the traffic situation description ontology.

Finally, issues concerning uncertainty handling are discussed in section 5.4 and a concept for basic uncertainty handling is provided.

5.1 An Ontology for the Description of Complex Intersection Situations

This section subdivides into five subsections to delineate the structure, design and usage of an ontology to describe complex traffic situations, like intersection situations. Subsection 5.1.1 first describes the basic idea and layout of such an ontology at the example of a very simple intersection. Subsection 5.1.2 then extends this ontology to arbitrarily large intersections and describes its layout and capabilities. Reasoning results and queries to this ontology are illustrated in the subsequent subsection 5.1.3. Subsection 5.1.4 further expands this complex intersection ontology to different applications and reasoning goals. Moreover, some logic reasoning is discussed for cases, when only partial knowledge is given in advance. Finally, subsection 5.1.5 provides an overview about the time consumption needed for loading, logical reasoning and querying this ontology.

5.1.1 Simple Intersection Description

During ontology engineering, a small ontology was first built to describe a simple intersection situation and perform reasoning about the simple right-before-left-rule. **Fig. 30** shows an illustration of this case with the types of objects involved. It does not show the complete and consistent situation, but the relevant ontological components. A picture showing a small real-world intersection of this type is shown in **Fig. 31 a)**. For description, notation and formalization of and reasoning on ontologies using description logic please refer to section 3.2.

The ontology contains the concepts Road, Crossing, Car and Sign with the subconcepts Yield-Sign and RightOfWaySign. It does not yet contain lanes. Based on these concepts, the roles are defined. Roads and intersections are related by connectedTo. To give a qualitative description of the position among roads, they are related with isRightOf. Automatically, the reasoner sets the inverse relation isLeftOf. Signs are assigned to both intersections and roads by isPart. isOn is used to relate cars to roads. These relations are set in advance, e.g. provided by sensor information. Depending on ontology engineering, some of the relations may also be reasoned instead. Finally, the goal is to reason the relations hasRightOfWay and hasToYield inversely. This traffic rule reasoning will be carried out by executing so-called ABox augmentation rules since role chains and other multivariable constructs are not supplied by the reasoner or DL [Baader and Nutt, 2009, Gries et al., 2010] (see section 3.2.4.3).

This ontology shows that simple intersections can be modeled effectively, allowing multi object situations to be created and interpreted with respect to a reasoning goal measure. The ontology was tested for correctness w. r. t. the right-of-way relations with all possible cases of approaching vehicles (none up to four approaching vehicles) and one departing vehicle. Different possible constellations of approaching vehicles were considered (but are limited in their number due to rotational symmetry). Note, that for a correct logical set of axioms and rules, the reasoning result has to be correct. See for example the results in [Hummel, 2009], which did not have any false positives or false negatives in the test cases.

Reasoning of an intersection ABox similar to **Fig. 30** takes less than 0.5 s on a quadcore CPU with 4x 2.4 GHz.

5.1.2 Complex Intersection and Road Network Ontology

5.1.2.1 Basic Structure

The goal is to build a basic ontology, capable of describing complex intersection situations with an arbitrary number of roads connected to the intersection, an arbitrary number of lanes on them and different constellations of traffic signs and lights. This ontology will be called Traffic Intersection Situation Description Ontology (TISDO). A picture of a very huge intersection at the Albertplatz in Dresden, Germany, with a large number of roads and lanes and several restrictions on and combinations of driving paths is shown in **Fig. 31 b)**. TISDO is designed to have the capability to describe intersections of arbitrary size including the depicted example.

The basis for relations and reasoning is formed by a taxonomy of concepts as depicted in **Fig. 32** further detailing the concepts of **Fig. 30**. Moreover, a role hierarchy is created as shown in **Fig. 33**. The idea of the partial taxonomy belonging to the concept RoadConnection is borrowed from [Hummel, 2009, Hummel et al., 2007].

Yet, reasoning is performed differently, aiming to reason about traffic rules for vehicles rather than geometry and ordering of lanes and markings in the scene.

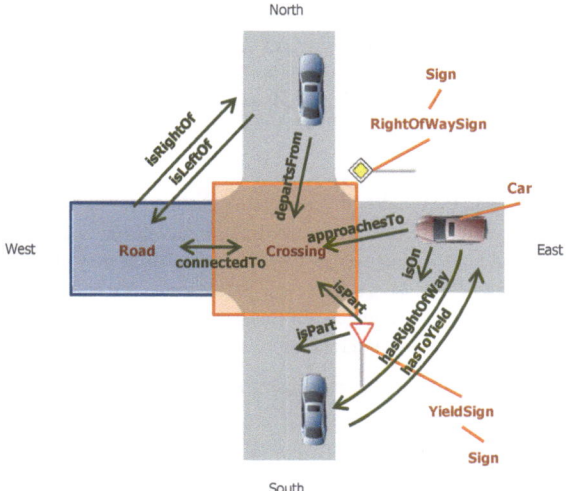

Fig. 30: Illustration of the elements of the simple intersection ontology.

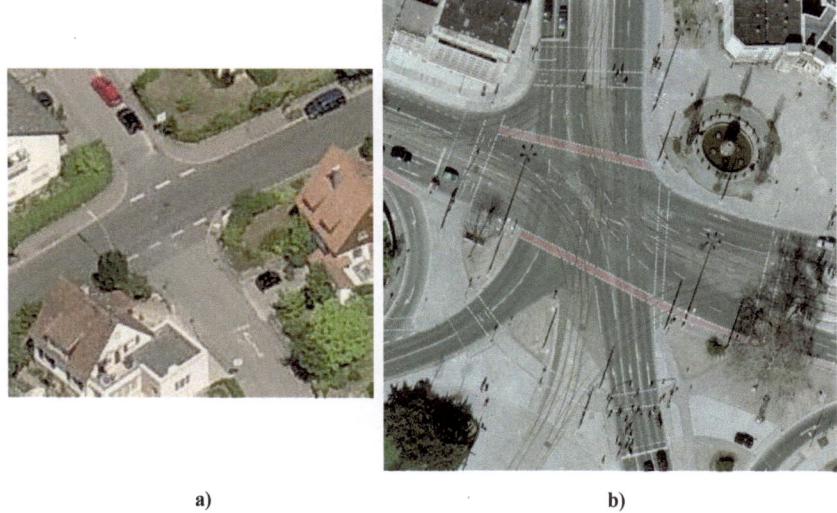

a)   b)

Fig. 31: **a)** Example of a simple suburban intersection. **b)** Example of a very complex intersection at the Albertplatz in Dresden, Germany. It consists of at least 5 roads (possibly dividable into up to 8 one-way roads) and 17 lanes. These numbers rise further, if bus roads and lanes are considered additionally.   (provided by Google Maps)

Most concepts within the same hierarchical layer are made disjoint. Disjointness is left out only for the concepts CrossingWTrafficLight and CrossingWTrafficSign. A crossing may both have traffic signs and lights and thus belong to both concepts at a time. Reasoning depends on whether existing traffic lights are set off, i.e. belonging to the concept TLaCr_Off.

5.1.2.2 Reasoning Objectives

The goal for reasoning and eventually for intelligent agents performing assistance functions in this example ontology is mainly to retrieve information about the relations hasTRuleRel (has traffic rule relation) and below. However, reasoning on an ABox may also create e.g. the relations approachesTo, departsFrom, isPart, hasEnteringRoadConnection and others. This creation of relations depends on the kind of information that is sent to the ontology in advance and, if enough information is provided to set a relation.

The focus of this example is to enable intelligent intersection DA agents to comprehend the traffic rule situation of vehicles.

Note, that this ontological design not only allows for reasoning with respect to one specific vehicle but all vehicles at an intersection. Moreover, a whole network of intersections can be created through connecting RoadConnection objects with Crossing objects[39].

In this way not only agents within a specific vehicle but also within a traffic light control or even a capacious traffic management system may be realized. This fact also promotes this ontology to be used for Car2Car (C2C) or Car2Infrastructure (C2I) information exchange.

Another capability of this ontology, resulting from consequently using its hierarchy of concepts and roles, is generic reasoning on higher level elements rather than each distinctive concept or role, e.g. roads and lanes separately. This allows, for example, obtaining traffic rule information, even if lane information is not given for each vehicle. For these vehicles, road information is used and less precise but yet correct traffic rule information is reasoned.

Several constructors and operations provided by the description logic dialect $\mathcal{ALCQHIr+(D)}$ introduced in **Table 1** and **Table 2** are required to formulate axioms related to infrastructure and traffic rules and the domain of ADAS, respectively. Examples of several axioms and rules are provided in sections 5.1.2.4 and 5.1.2.5.

Recall that the $\mathcal{ALCQHIr+(D)}$ dialect includes atomic negation, concept intersection, existential and universal restrictions, complex concept negation, qualified cardinal restrictions, role hierarchies, inverse roles, transitive roles and concrete domains (attribute data types).

Concept and role hierarchies were motivated in the previous paragraphs. Atomic negation enables axioms such as "an entering lane is *not* an exiting lane" (LaneEntering ⊑ ¬ LaneExiting). Concept intersection is required to create complex concepts such as "a car *that is* on a lane" (Car ⊓ isOn.Lane). (Full) existential restrictions are required for axioms such as "For each road *there is* a (at least one) lane" (Road ⊑ ∃ hasPart.Lane) where Lane is the qualified domain ("full" meaning all concepts are allowed for restriction). Universal (Value) restrictions are

---

[39] Actually, the classification to the concept Crossing does not necessarily have to be set, but is reasoned automatically, if at least three roads are connected to the object. Traffic rule reasoning depends on Crossing objects.

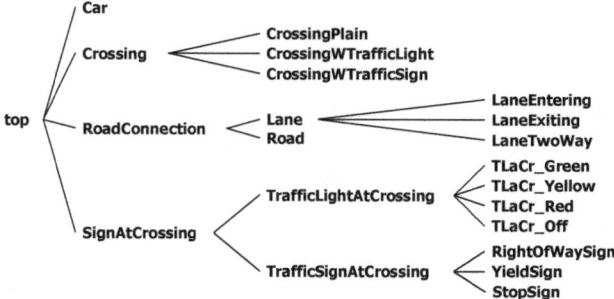

Fig. 32: Taxonomy of the complex intersection ontology (as plotted by RacerPorter) as an excerpt of a developed generic traffic situation taxonomy.

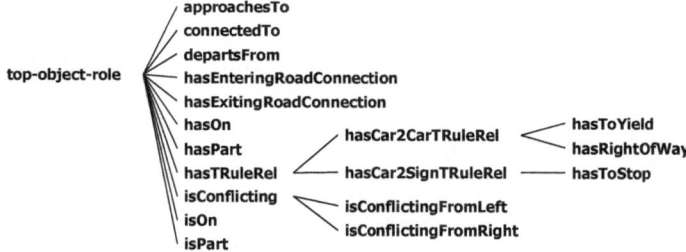

Fig. 33: Role hierarchy of the intersection ontology (as plotted by RacerPorter).

used for statements such as "Road connections are *only* connected to road connections or crossings" (RoadConnection ⊑ ∀ connectedTo.(RoadConnection ⊔ Crossing)). (Qualified) Cardinal restrictions allow axioms such as "A car is on *exactly one* road / is on *at most* one lane / has *at least* one exiting road connection" (Car ⊑ ∃$_{=1}$ isOn.Road / Car ⊑ ∃$_{\leq1}$ isOn.Lane / ∃$_{\geq1}$ hasExitingRoadConnection.RoadConnection).

Complex concept negation enables to negate combinations of the above stated complex concept constructors.

Inverse roles are very useful for automatic reasoning of some relations and extensively used in the ontology. In this way, relations such as isPart, isLeftOf, connectedTo are automatically reasoned from hasPart, isRightOf, connectedTo, for example. Transitive roles, however, were not used due to conflicts arising when using cycles of relations such as isRightOf between roads at crossings or when using number restrictions such as "exactly one". Transitive roles are automatically created with hardly any supervision as they depend on the specifics of an ABox rather than TBox axioms, which can hardly be known during knowledge engineering. For example, a relation isRightOf might be set only once for each object, but in the case of a transitive role, the condition "exactly one" would hold only if there was not another relation isRightOf at the object laying left. Otherwise, further relations isRightOf would be automatically created.

Finally, concrete domains are necessary to store attribute values and create according axioms using the attributes such as angles or distances.

Role chains and other complex role axioms such as inclusion, disjointness and others would be very useful for the formulation of e.g. traffic rules (as explained in section 5.1.2.5) and to check their satisfiability. However, they are not provided by reasoners or computationally too expensive (see section 3.2.4.3) and rules had to be used.

Nominals were not needed, as they are easily replaceable by concrete domains.

For a more powerful generic situation description, abductive logic (creating instances through axioms) and probabilistic logic reasoning would be the next step to focus on. Unfortunately, these methods are still under advance research and not yet sufficiently expressive, available to use or computationally too expensive (see sections 3.2.5, 5.4).

For the generic situation description provided with this thesis, the description logic dialect $\mathcal{ALCQHIr+(D)}$- used for ontology engineering is sufficiently expressive and together with the RacerPro reasoner efficient to use.

5.1.2.3 Module Based Ontology

Ontologies allow exchanging ontology parts or loading additional ones. In this way, lean ontologies can be made more complex if needed. E.g. detailed road geometry for extended assistance functions could be added with the ontology from [Hummel, 2009, Hummel et al., 2007] (a brief explanation is provided in section 2.2.6). These modules are formed by different TBoxes that are merged to a complete TBox. To include this kind of module-based layout, TISDO is structured into the following four TBoxes:

- The *Taxonomy* TBox entails concept implications and closure axioms to build the concept hierarchy. For the intersection ontology, only a part of a large driver assistance and traffic situation taxonomy is loaded and shown in **Fig. 32**. That large taxonomy developed for this thesis contains noticeably more than 100 concepts and is too large to be illustrated in this thesis. The part of the taxonomy used and illustrated here, might be easily exchanged with bigger parts or even the complete taxonomy to allow additional modeling and reasoning.

- The *RoadNetwork* TBox accounts for the elements used to connect intersections with roads, lanes and some lean geometric description. Instead of using the isRightOf, isLeftOf relations, an angular attribute hasAngleAtCrossing is now used to encode the position of roads and lanes w. r. t. the intersection. This might e.g. be replaced by the ontology in [Hummel, 2009, Hummel et al., 2007] if needed.

- The *Crossing* TBox contains additional constraints about how traffic signs have to be related and axioms about how the type of crossing (see subconcepts in **Fig. 32**) is determined.

- Additional constraints for vehicles in the network are given in the *Car* TBox e.g. a vehicle can only be on one road / lane at a time. It constrains the driving direction with hasEnteringRoadConnection only being related to LaneEntering or LaneTwoWay (but not LaneExiting). These constraints keep sensor and test case consistency.

- The *Rules* module, which is no actual TBox, contains all ABox augmentation rules to be executed after reasoning. The module can be further split to allow exchangeability e.g. for considering regional differences in traffic rules.

Note, that ontological constraints do not enable extensive automatic building of an ABox due to the OWA, but disallow inconsistent ABoxes to be built. They define the semantic framework for how to build an ABox and for its meaning.

5.1.2.4 Description Logic Axioms

The four TBoxes contain the background knowledge about an intersection situation. The structure and content will be illustrated on several example axioms and rules. To apply the appropriate traffic rules, depending on the situation, and determine e.g. right-of-way relations of vehicles, it is necessary to know the constellation of roads or lanes at the intersection. This is encoded with angles of roads and lanes in a polar intersection coordinate system, assuming that an angular representation is also inherent to human comprehension of an intersection layout. Unless trajectory planning for a specific destination is necessary, humans have a rough map of the roads' positions in mind. With these angles (or the image using them) in mind, reasoning about relative positions like "the road between two others" is easily possible, without implementing loop-until directives[40].

To enable generic reasoning with either roads or lanes (depending on the information detail available after sensory perception), the abstract concept RoadConnection is used according to [Hummel, 2009, Hummel et al., 2007]. "Abstract" means, this concept will not be instantiated. With the taxonomy, Road and Lane instances are classified belonging to this concept RoadConnection. The advantage lies in the application of traffic rules on this concept rather than on roads or lanes distinctively. This saves engineering effort, amount of axioms and enables reasoning, even if lane information is not available for some roads. The following excerpt gives some example axioms of the *RoadNetwork* TBox:

```
T_RoadNetwork ⊃ {
        Crossing ≡ ∃≥3 connectedTo.Road
        RoadConnection ⊑ ∃=1 connectedTo.Crossing
        RoadConnection ⊑ ∃≤1 connectedTo.RoadConnection
        RoadConnection ⊑ (hasAngleAtCrossing ≥ 0) ⊓ (hasAngleAtCrossing < 360)
        Road ⊑ ∃ hasPart.Lane   } .
```

This is the TBox providing knowledge for vehicles:

---

[40] Such loop-until directives would be necessary, if isRightOf / isLeftOf would still be used for arbitrary intersections.

$T_{Car} \supset \{$
>   Car $\sqsubseteq \exists_{=1}$ isOn.Road
>   Car $\sqsubseteq \exists_{\leq1}$ isOn.Lane
>   Car $\sqsubseteq (\exists$ approachesTo.Crossing
> $\sqcap \exists$ hasEnteringRoadConnection.(RoadConnection $\sqcap \neg$ LaneExiting)
> $\sqcap \exists$ hasExitingRoadConnection.(RoadConnection $\sqcap \neg$ LaneEntering))
> $\sqcup (\exists$ departsFrom.Crossing) $\} .$

The *Car* TBox is especially suited to check for sensor errors like false positioning of vehicles, e.g. onto two lanes or on an exiting lane, when the vehicle is entering the intersection.

These axioms describe how a traffic intersection situation is set up in the developed TISDO and define its semantics.

5.1.2.5 Rules and Underpinnings due to Lack of Role Chains

To reason about relations, rules are needed. They fire, that is, their rule consequence is set, if their conditions (antecedence) hold, and set the desired assertions. Simple examples of the extensive set of partially very complex rules will be given here. Since role chains are not applicable (see section 3.2.4.3), a rule is created, inferring that a car that is on a lane is also on the road, the lane itself is part of:

isOn(?car, ?ln) $\wedge$ isPart(?ln,?rd) $\wedge$ Car(?car) $\wedge$ Lane(?ln) $\wedge$ Road(?rd)
> $\rightarrow$ isOn(?car,?rd) .

The according axiom formulated with a role chain would be

Car $\sqcap$ isOn.Lane $\circ$ isPart.Road $\sqsubseteq$ isOn.Road .

Because of the OWA an intersection cannot be classified as CrossingPlain (without traffic signs or lights), if no definite information is provided, e.g. by the sensors. The goal of this example ontology is to still infer the simple right-of-way traffic rules. This kind of reasoning is enabled by using *negation as failure* (neg operator). The according rule is

Crossing(?cr) $\wedge$ neg(CrossingWTrafficLight(?cr)) $\wedge$ neg(CrossingWTrafficSign(?cr))
> $\rightarrow$ CrossingPlain(?cr) .

With a strict application of OWA, sensor information like $\neg$ isPart.Sign is required to be given and the same result will be produced with axioms of the *Crossing* TBox.

Last but not least, a simple right-of-way traffic rule will be shown. Several rules of this kind exist in the ontology.

CrossingPlain(?cr) $\wedge$ Car(?c1) $\wedge$ Car(?c2)
> $\wedge$ approachesTo(?c1,?cr) $\wedge$ approachesTo(?c2,?cr)
> $\wedge$ isConflictingFromRight(?c2,?c1) $\wedge$ neg(same-as(?c1,?c2))
> $\rightarrow$ hasRightOfWay(?c2,?c1) .

Preceding these traffic rules, there are rules reasoning about whether cars are conflicting each other and whether cars have to obey some traffic signs that influence their state of right-of-way.

5.1.3 Reasoning Results of an Intersection Example

5.1.3.1 Complex Intersection Example

To show the capabilities of the introduced intersection ontology, some results will be discussed on a huge example intersection with 5 roads, 11 lanes and 6 cars, as depicted in **Fig. 34**. For the application of traffic rules, an exemplary configuration of right-of-way and yield signs will be used.

The legend indicates the ability to incorporate traffic lights as well. Each road is mentioned with its geometric angle attached. **Fig. 34** plots an image of the knowledge incorporated in the ontological description. The thin, black arrows define the lanes on roads and their allowed driving direction, entering or exiting the intersection. Blue circles depict cars and the thick, red arrows their desired driving path. However, the path is stored in the ontology plainly by an entering lane and an exiting lane for each car. Note, that this means though that uncertainty about a cars destination cannot be handled directly. These cases would have to be considered separately with the TISDO approach. Some uncertainty handling will be described later in section 5.4.

TIDSO was tested on many other constellations of roads, lanes, with or without traffic signs or lights and approaching or departing vehicles, coming from different directions. It always yielded correct reasoning results.

5.1.3.2 Resulting Ontology

For the example of **Fig. 34**, the ontology reasons right-of-way of cars 1 and 3 over cars 2, 4 and 5 due to their respective right-of-way sign. However, car 3 has to yield car 1, as it desires to make a left turn. The ontology reasons car 1 to be conflicting car 3 from right and thus applies the right-before-left rule.

As cars 2, 4 and 5 all have an according yield sign, they have to regard the right-before-left rule among them. In this way, car 2 has right-of-way over car 5 and car 5 over car 4. All other combinations do not have any additional traffic rule relation, since e.g. car 2 and car 4 could drive to their destination without conflict. Car 6 is leaving the intersection and thus not considered.

The resulting ABox after reasoning is large, even for this lean ontology example containing 29 individuals, 44 concept assertions and 75 role assertions. Some relations to other objects reasoned for car 5 will only be exemplary discussed.

From car 5 to car 2, there exist relations isConflicting, isConflictingFromLeft, hasTRuleRel, hasCar2CarTRuleRel and hasToYield. isConflictingFromLeft and hasToYield are reasoned with rules, the others are then inferred with the role hierarchy given in **Fig. 33**. In contrast, there are fewer relations from car 5 to car 3, namely hasTRuleRel, hasCar2CarTRuleRel and hasToYield. Their paths passing the intersection are not necessarily conflicting, so the relations about conflicts are not set (but may still exist due to the open world assumption). The constellation of traffic signs still accounts for the right-of-way relations between these two cars.

Corresponding relations exist between all other pairs of cars.

Treating intersections with traffic lights and stop signs, cars may receive relations hasToStop to the according element. Right-of-way rules for traffic lights are treated similarly as with traffic signs.

Besides the traffic rule relations, car 5 has the relations isOn to road 5, hasEnteringRoadConnection and isOn to lane 1 on road 5, hasExitingRoadConnection to lane 2 of road 2 and approachesTo to the intersection.

5.1.3.3 Query Examples

Intelligent agents carrying out DA functions are aimed to make use of the situation description knowledge base TISDO, introduced with this thesis. The knowledge base is able to answer questions about the situation (see ABox inference services above).

These are a few example queries to the ontology that might be posed by DA agents to retrieve information for their functions:

- Query 1: Retrieve all pairs of instances with the relation isConflictingFromRight.
  Answer: (car 5, car 4), (car 2, car 5), (car 1, car 3), (car 3, car 2)
- Query 2: Retrieve all pairs of instances with the relation hasRightOfWay.
  Answer: (car 5, car 4), (car 2, car 5), (car 1, car 3), (car 1, car 5), (car 1, car 2), (car 1, car 4), (car 3, car 2), (car 3, car 5), (car 3, car 4)
- Query 3: Retrieve all instances with the relation hasToYield coming from car 3.
  Answer: (car 1)
- Query 4: Retrieve all instances with the relation hasToYield coming from car 1.
  Answer: none
- Query 5: Retrieve all instances with the relation hasExitingRoadConnection to lane 3 of road 4..
  Answer: (car 1), (car 3)

With this kind of queries, a DA agent might for instance find out, whether there are any vehicles that have right-of-way over the own vehicle. Subsequently, it can use that information for a situation dependent prediction of driving trajectories or maybe coverage probability densities. In the case of a violation or misjudgment by the driver who is not braking appropriately, an intersection warning assistant could initiate suitable actions, like warnings or autonomous braking maneuvers.

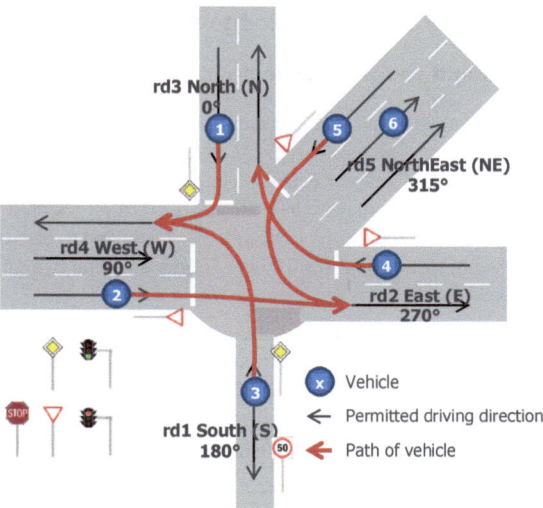

Fig. 34: Example of a complex intersection with 5 roads, 11 lanes (entering, exiting or two-way), 6 vehicles and traffic signs (possible signs see legend).

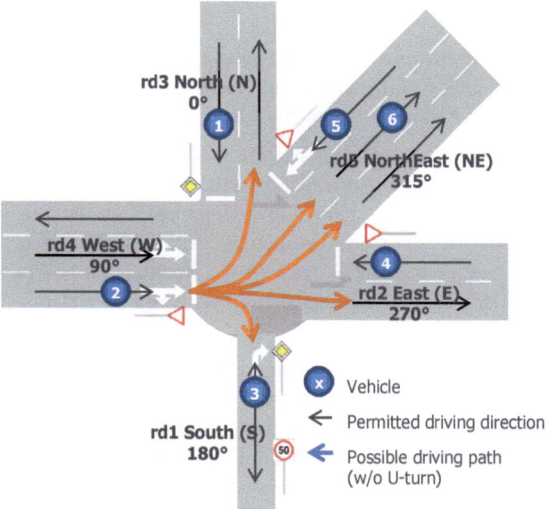

Fig. 35: Example of the complex intersection situation considering all possible driving paths (depicted by thick arrows for car 2 coming from road 4 West).

5.1.4 Expansions for Partial Knowledge Modeling

5.1.4.1 Worst Case Right-of-Way Reasoning

The hitherto introduced ontology provided reasoning about right-of-way relations between vehicles, when the exact driving destination of a vehicle is known in advance. However, this case may rarely exist due to ambiguity or occlusion of e.g. direction indicators, lane markings or even irregular behavior. Thus, it may be advantageous to reason about all possible driving paths and gain knowledge about the "worst case", that is, a vehicle has to yield another one in any one situation. As well, it can be shown, if there is only one kind of right-of-way relation to other cars, implying certainty.

Due to the OWA, *worst case* reasoning can be achieved by applying a rule setting the hasExitingRoadConnection relation for a vehicle to all exiting lanes, which are not part of the road it is currently on, as partially shown in **Fig. 35** for car 2. Reasoning using the inferred isConflicting relations is then performed for all combinations. In this way, a vehicle may receive the relations hasToYield as well as hasRightOfWay simultaneously, indicating both eventualities.

If the path of one vehicle is known, the possible destinations may yet be set for all others. If this vehicle e.g. plans to perform a direct right turn, right-of-way over all other cars would be reasoned correctly.

5.1.4.2 Modeling Allowed Driving Paths

Thinking further, the idea of *worst case* reasoning can be used to reason with restrictions on driving paths, meaning, only specific target exiting lanes are permitted for each entering lane. The concept is depicted in **Fig. 36** by the arrows where restrictions occur.

To allow this kind of reasoning, another relation hasAllowedExitingRoadConnection between lanes is established that is used analogically to hasExitingRoadConnection. The relation may be set when sourcing navigation data or retrieving environmental sensor data of e.g. lane markings or road signs, as shown in **Fig. 36**. Based on the additional relation, conflicts of allowed paths are inferred, if they are relevant for cars on the according lanes. The resulting conflict and right-of-way relations may be interpreted similarly to the "worst case" handling shown above.

The result, though, may be a lot more informative. In the situation depicted in **Fig. 36**, e.g. car 3 is only permitted to turn right. Consequently, right-of-way will be reasoned over all other cars that may have the same exiting lane as their destination, i.e. cars 1 and 2. Furthermore – due to the constellation of traffic signs – right-of-way is inferred over cars 4 and 5 without any conflict relations set. In this way, car 3 definitely has right-of-way in any case in contrast to "worst case" reasoning where car 3 may both have right-of-way and have to yield.

The two implementations for reasoning with all possibilities as well as restricted driving paths show the convenience of using ontologies for this kind of abstract traffic situation modeling and reasoning of e.g. traffic rules. Having formed the basic ontology, it can relatively easily be adapted to further applications and reasoning tasks without changing the overall structure.

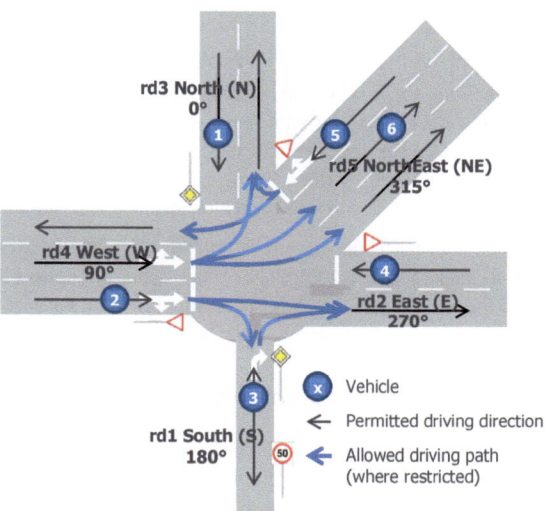

Fig. 36: Example of the complex intersection situation with predefined restrictions on allowed paths (depicted by thick arrows where restrictions apply).

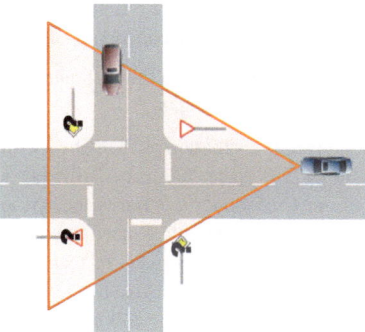

Fig. 37: Exemplary intersection illustrating uncertainty of right-of-way traffic signs not in the scope of a vehicles environmental sensors (field of view).

5.1.4.3 Reasoning for Partial Knowledge of Traffic Signs

Besides the handling of uncertainty of driving paths, as described above, uncertainty of right-of-way signs at the intersection, not in the field of view of sensors, are an open issue as exemplary shown in **Fig. 37**.

Actually, this case may only occur with intersections, which are connecting 3 or 4 roads. For more roads, the constellation of traffic signs cannot be determined by just one right-of-way sign and is clarified at the intersection by supplementary signs that indicate the main road with right-of-way. This is also done for turning main roads.

To take care of the remaining cases with only one traffic sign detected and the others to be reasoned, the ontology has been enhanced with appropriate rules. First it is helpful to determine adjacent roads, wherefore the relations isRightOf and its inverse isLeftOf as well as the parent role isAdjacentTo are reasoned by rules. These relations can then be used to reason the unknown traffic sign classes YieldSign or RightOfWaySign for the remaining traffic signs.

Abduction with automatic instance generation is still an ongoing research field (see e.g. [Gries et al., 2010, Kate and Mooney, 2009]). For this reason, all signs still have to be set completely in advance as belonging to the class TrafficSignAtCrossing for each road at an intersection. Reasoning then accounts for allocation of the right traffic sign classes.

5.1.5 Time Consumption for Reasoning and Implementation Issues for DAS

The goal of this thesis is not only to create an ontology capable of describing complex traffic situations, e.g. at intersections as proposed here, but also to enable reasoning that is close to real-time and may soon be implemented as such.

Reasoning of the example of a complex intersection with 5 roads, 11 lanes and 6 cars takes about $0.97\ s$ without any road signs and with $1.17\ s$ slightly more, when traffic signs and / or lights are involved. Reasoning was executed on a quadcore CPU with 4x 2.4 GHz.

The simple standard intersection example with 4 roads, 8 lanes and 4 cars takes about $0.66\ s$ without signs and $0.77\ s$ with traffic signs or lights.

For an understanding of an intersection situation as described with this thesis, reasoning time of less than 0.5 s on a vehicle ECU is suspected to be sufficient for real-time usage. This is, because the ontologically reasoned information is not suspected to change highly dynamically. Roads, lanes, signs and cars, after they are recognized, are not expected to suddenly disappear[41]. Consequently the gap of this thesis' approach to real-time applicability is less than a dimension in time consumption or even already close to the assumed real-time applicability. Real-time capability for some exemplary functions will be proven in sections 5.2 and 5.3.

In contrast to the introduced ontology, the urban intersection scene understanding ontology in [Hummel, 2009, Hummel et al., 2007] carries out extensive reasoning on geometry and constraints for the structural elements of an intersection. Consequently reasoning there may take up to hours in comparison to approximately one second for the herewith introduced approach.

However, reasoning about road geometry or other issues may sometimes be not as time critical as the description of a situation the vehicle is currently in. As a solution, applying multi-layer ontologies can be thought of, to separate different layers of abstraction and different time criticality levels (please refer to next section, particularly subsection 5.2.3).

---

[41] Existence probabilities can be treated separately and will be handled in section 5.4.

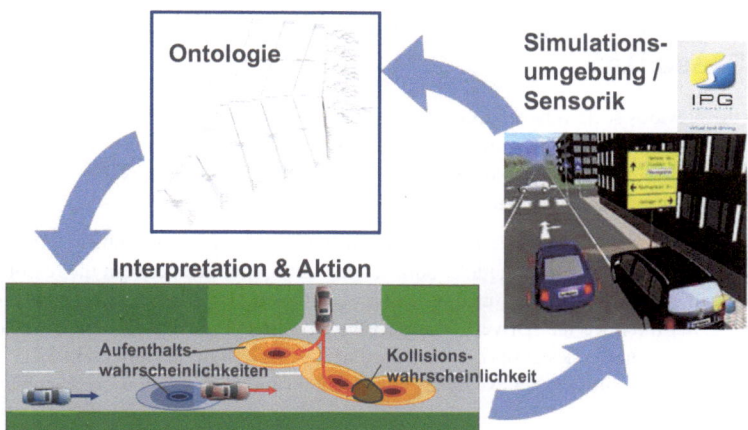

Fig. 38: Flow of operation of a knowledge-based DAS using a vehicle simulation environment to simulate vehicle dynamics, onboard sensors and actors.

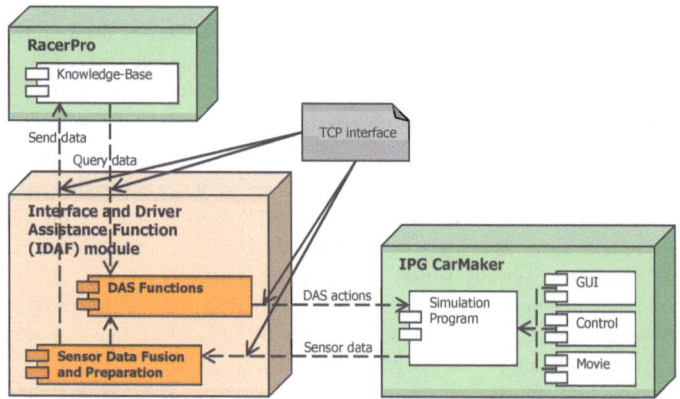

Fig. 39: Modules and interaction of the asynchronous real-time framework containing a simulation environment, a knowledge representation environment and the introduced Interface and Driver Assistance Functions (IDAF) module.

The low time consumption was reached mainly by proper formulation of axioms, omitting redundant formulations included in different axioms, omitting axioms with an overhead of functionality not needed for the content addressed with the ontology and to a high extent by regrouping or splitting augmentation rules and reordering their conjunctive and disjunctive terms.

5.2 Asynchronous Real-Time Framework for Knowledge-Based DAS

While the previous section introduced the ontology TISDO to effectively describe traffic situations at complex intersections, this section deals with efficiently implementing this ontology in combination with a reasoner for prototype real-time DAS functions within a realistic 3D vehicle dynamics simulation environment.

First, subsection 5.2.1 introduces the concept of ontology-based DAS function simulation. Subsection 5.2 then presents a module, both serving as an interface between the reasoner and a simulation environment and containing sensor data fusion of sensor data and DAS function routines – the Interface and Driver Assistance Function module (IDAF). This module may also feed several ontologies at once, which is subject of discussion in subsection 5.2.3. Finally, subsection 5.2.4 presents an exemplary intersection simulation to show the capabilities of the IDAF module.

5.2.1 Simulation of Knowledge-Based Driver Assistance Systems

Evaluation of the applicability of the ontology proposed in this thesis is done using a 3D vehicle dynamics simulation environment. This not only aims to simulate real world ECU implementation. It also allows modeling defined traffic situations for testing and providing any sensor information needed that may not yet robustly be available from vehicle onboard sensors.

The basic flow of operation when using an ontology, based on a vehicle dynamics simulation, is shown in **Fig. 38**. The vehicle dynamics simulation environment is mainly used for testing. Vehicle dynamics are simulated including the onboard sensors, to capture the vehicle environment and vehicle state. Actors are simulated as well, to perform DAS actions that influence vehicle behavior. Cyclically, sensor data is provided by the simulation environment and is transformed into an ontology. Queries to the ontology then provide information for further interpretation and DAS function calculation (decision making). DAS functions send actuatory commands to the simulation environment to perform DAS actions.

5.2.2 Asynchronous Real-Time Framework

5.2.2.1 Architecture and Communication

To implement knowledge-bases for intersection situation description, an asynchronous real-time framework was created. The basic structure is depicted in **Fig. 39**. Development and testing was performed with the vehicle dynamics simulation environment (IPG CarMaker [IPG Automotive, 2011]). Thus, the vehicle, its environment and especially the sensory and actory components are replaced by the simulation. The RacerPro module contains the knowledge-base and performs reasoning on it. Both modules provide a TCP-interface, such that other software may communicate with them.

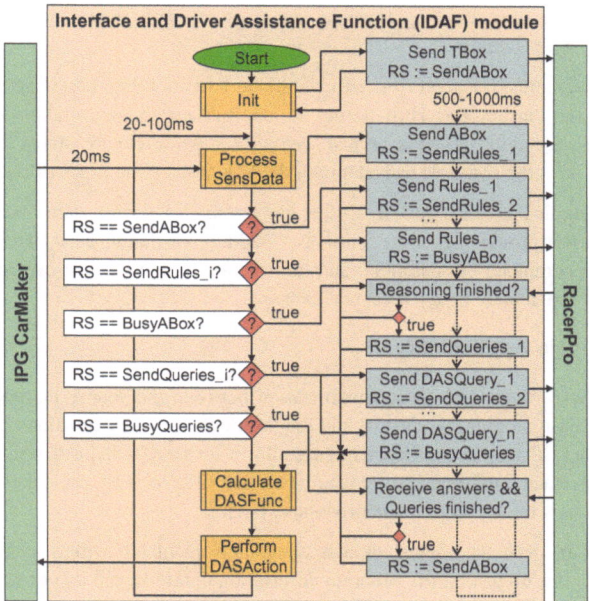

Fig. 40: Flowchart of routines and interaction of the IDAF module. (RS: RacerPro Status)

As the simulation module only represents the sensory and actuatory part of a DAS system, DAS functions are executed in another module, the new Interface and Driver Assistance Functions (IDAF) module. IDAF not only accounts for calculation of DAS functions, but also establishes and ensures communication between these modules. It consists of two major components, the DAS functions module and the Sensor Data Fusion (SDF) module. The latter prepares sensory information to be sent to the knowledge-base.

All three modules run in parallel and exchange information coordinated by the IDAF module. Sensor data is updated very frequently at a cycle time of about 20 *ms*. Reasoning on an ABox of the ontology within the RacerPro module takes much longer, ranging from about 0.5 *s* up to 2 *s*, depending on the reasoning mode and the situation. Therefore, the framework was developed to be asynchronously running and communicating between the modules. Its cycle time varies between 20 *ms* and 100 *ms*.

5.2.2.2 Flow of Operation

The flow of operation of the IDAF module is shown in **Fig. 40**. Colors of the modules correspond to those illustrated in **Fig. 39**. After initializing the TCP-connections and modules and sending the TBoxes to RacerPro, the main routine of the IDAF module runs a series of subroutines in each cycle. In each cycle the sensor data is evaluated and processed for further usage in the ontology and DAS functions. The latter are calculated and, if appropriate, DAS actions are performed in each cycle. However, this does not apply for the ontology routines for communication with RacerPro.

In each cycle the status of RacerPro is checked by an internal token flag (RacerPro satus RS). Depending on the current RacerPro status, the appropriate action (sending ABox or rules, sending queries, receiving answers or just waiting another cycle for status updates) is performed. In this way, the software can deal with the long processing times of the ontology.

Using the token flag enables fast, real-time processing of sensor data and DAS actions, whereas the slower processing of the ontology (loading, reasoning, querying etc.) runs asynchronously and completely in parallel[42] and information within the fast cycle routine is only updated once it is available by the reasoner.

5.2.2.3 In-vehicle Applicability

In this simulation framework, the communication between the sensor module (simulation), the knowledge-base module (RacerPro) and the interface and DAS functions module (IDAF) is realized with TCP interfaces. This structure allows for easy transformation into in-vehicle implementation. Each module may be executed on one or several ECUs, communicating with each other via CAN bus or other automotive bus systems.

In the simulation, necessary object data is completely available within a vehicles field of view. The focus lies on testing the situation description ontology and its general applicability. Sensor information may thus originate from arbitrary sensors which allows for various applications using different sensor setups.

5.2.3 Multilevel-Structure of Ontology Functionality for Online Traffic Situation Description

Before getting to different applications and implementations of the IDAF module in subsection 5.2.4 and section 5.3, one of its capabilities shall be discussed with respect to different types of ontologies concerning their expressiveness.

The IDAF module is designed to be capable of communicating with a virtually arbitrary number of ontologies and reasoners. On one hand, it can communicate with several reasoners that may even run on several ECUs. On the other hand, it may as well run one TBox on one reasoner and send and query several ABoxes running on the same reasoner. In this way, different instantiations of the same ontology (TBox background knowledge) may be reasoned out. This is especially helpful to reason about different possibilities or possible worlds, respectively. Handling such worlds is object of uncertainty handling in section 5.4.

The capability of handling several ECUs, reasoners and ABoxes can be used in different ways explained in the following subsections.

---

[42] The entire simulation was running on several separate CPUs of the desktop: one running the IDAF calculations, one running the RacerPro reasoner, another one running the simulation environment and a fourth one running the graphical environment (besides another separate GPU).

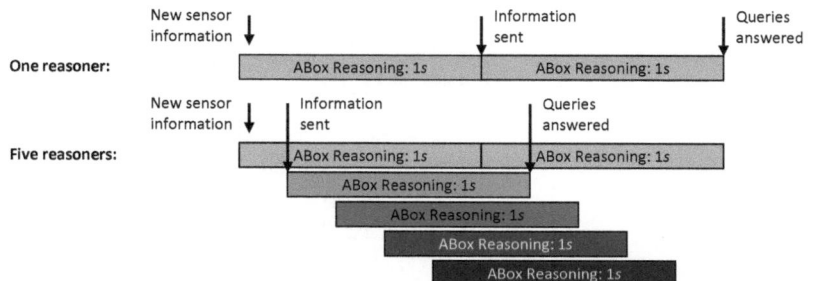

Fig. 41: Using several reasoners simultaneously to reason on the same TBox may significantly reduce query answering delays for DAS functions (in the figure a reasoning time of 1s is assumed).

Running the Same Ontology in Parallel

In general, the maximum time, a DAS function can respond to new information (e.g. a newly sensed intersection), is twice the time that is needed from the perception of new information, sending it to the ontology and waiting for the answer to be acquired from the reasoner. That is, because the reasoner may have just started running reasoning and a whole reasoning cycle has to be waited to send the new information. An illustration of this effect is shown by **Fig. 41**. Note, that the design of the IDAF module yet allows using up-to-date dynamic data for DAS functions. These DAS functions are able to continue performing their calculations and actions during logic reasoning on the ontology.

Because the IDAF module is able to handle multiple reasoners, the same ontology may also be running on several reasoners. In this way, new information can be send to the next reasoner, as soon as it finishes reasoning and answering queries. Let us assume, reasoning time is 1 s. The maximum time for a DAS function to act is 2 s accordingly. But if, for example, 5 reasoners are running in parallel, this maximum response time could be reduced to 1.2 s.

Running Ontologies of Different Functionality

Fig. 42 shows, where different types of ontologies can be positioned in a plot over their range and capability of modeled domain knowledge and their amount of reasoning. The *"functional capability"* on the horizontal axis specifies how much knowledge about a domain can be modeled with an ontology, how generic, extendable and expressive an ontology may be. Low functional capability means, an ontology is designed for a very specific part of a domain. With higher functional capability, there is a tendency to longer reasoning time on axioms and rules. This is due to more concepts, a bigger taxonomy, more axioms and probably more individuals and relations to reason about.

In contrast, the vertical axis represents the amount of reasoning or *"logical capability"* of the ontology. It specifies how much high-level reasoning is necessary for the ontology or, how many rather complex axioms and augmentation rules are contained in the ontology. High logical capability also stands for comparably long reasoning times.

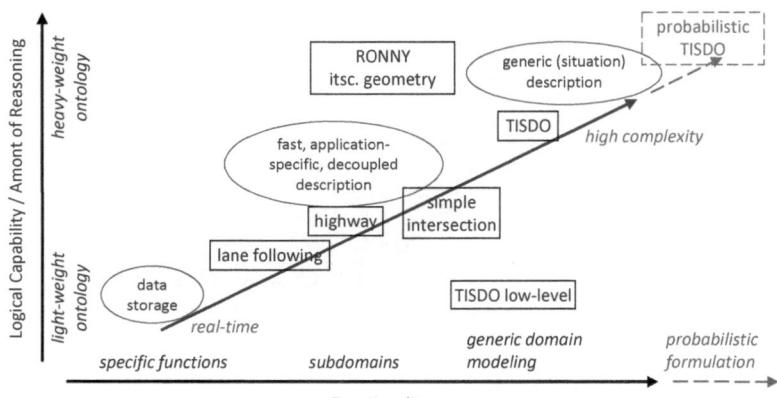

Fig. 42: Multilevel-structure of ontologies by their functionality and reasoning complexity. Circles roughly represent groups of ontological realizations whereas rectangles give examples. For TISDO refer to sections 5.1.2 and following, for the simple intersection ontology to 5.1.1 and for RONNY to [Hummel, 2009]. TISDO low-level refers to TISDO with omitted high-level reasoning (rules and complex axioms) that can be used for real-time storage and fast low-level reasoning.

As a consequence, the ordering of ontologies as in **Fig. 42** may also serve to develop ontologies with different functional capability and logical capability, depending on the time criticality needed and the target use. Here, this is referred to as a multi-level ontology structure or 2-dimensional ontology structure with the dimensions as presented.

Along one dimension, different ontologies serving different target applications, preferably different groups of DAS functions, may be developed that are leaner and, hence, faster in reasoning. The IDAF module allows these ontologies to run in parallel, e.g. on different ECUs or reasoners, respectively. This reduces reasoning and response times on situation changes for implemented DAS functions.

Section 5.1.4 introduced several modes to run TISDO to reason about different assumptions on traffic rule compliance. With IDAF, these modes (TISDO variants) can be run in parallel and information queried and used depending on the situation.

Running Ontologies with Different Amount of Reasoning

Along the other dimension, ontologies with diverse logical capability may be developed to facilitate fast accessibility on time critical information with most reasoning omitted (light-weight ontologies) as well as to provide high-level logical reasoning resulting in complex information (heavy-weight ontologies). Time critical information is such as dynamic data and situative road structure and geometry, where only few logical reasoning is necessary. Complex information may include reasoning results about traffic rules and legislation, infrastructure constraints and consistency of situations. While the TISDO is a heavy-weight ontology, including extensive reasoning about the latter aspects, **Fig. 42** also contains a "TISDO low-level" which is a proposal for a light-weight TISDO with most of its reasoning abandoned.

This ontology would contain only some very fast classifications using the taxonomy and, if needed, some fast, direct reasoning about relations, e.g. about preceding vehicles or vehicles on the ego lane.

Another example for a low-level ontology including highly dynamic data is the ontology for trajectory simulation by [Durak et al., 2006].

5.2.4 Intersection Simulation

5.2.4.1 Sample Intersection Assistance System

Intersection assistance is especially useful in situations with driver information overload, large intersection complexity, high speed of traffic participants, occlusion of relevant objects or driver distraction. To perform this kind of assistance, the system has to *know* about the intersection itself. Moreover, it has to know about connected roads, lanes and allowed driving paths where required. It has to know about relevant road signs like traffic lights and traffic signs and the current traffic participants associated with the situation. Furthermore, it has to be capable to even reason about right-of-way situations, constellations of traffic signs and lights and possible driving paths.

Fig. 43 depicts a simple intersection situation, containing some of the mentioned elements. It shows the principle of building an exemplary intersection collision avoidance assistance with right-of-way information and warning. This assistance example shows the general capability of TISDO to be used for assistance functions in real-time. It is based on several Time To Intersection (TTI) calculations and an avoidance acceleration. These measures are permanently recalculated with updates of dynamic and knowledge data. A potential collision is detected using the TTIs, when vehicles are expected to be in a certain area (orange) at the same time and subsequently, the necessary collision avoidance acceleration is determined for DAS functions. The underlying time measures, such as TTIs are calculated quasi-statically, hence, including current velocities and accelerations. More complex algorithms exist for state of the art collision avoidance systems. However, to show general applicability of the ontology, this complexity is not necessary.

5.2.4.2 Real-Time Simulation Results

For this thesis, the real-time framework was first tested on a standard intersection with 4 roads, 8 lanes and right-of-way signs, according to the illustration in **Fig. 43**. Snapshots of the simulation for an approach of the silver ego vehicle that has to yield another approaching red vehicle are shown and compared in **Fig. 44**. Both vehicles are approaching the intersection at a speed of 50 *km/h*. The left column shows situation progress with DAS functions deactivated, resulting in a collision of both vehicles. The right column, with activated DAS functions, shows a warning (yellow pyramid above the red vehicle as potential collision object at t = - 1.98 *s*) of an upcoming collision, which can only be avoided by an uncomfortable braking maneuver (warning condition). It is followed by an autonomous emergency braking maneuver

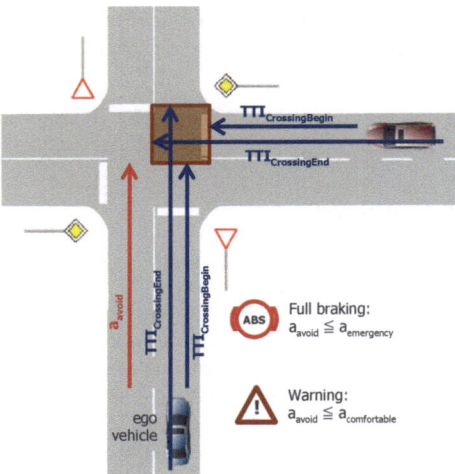

Fig. 43: Intersection assistance illustration. For two approaching vehicles an upcoming collision is detected based on different Time To Intersection (TTI) calculations using driving paths and dynamic data. Driver assistance is then performed based on avoidance acceleration evaluations.

(additional red pyramid above the red vehicle and green brake force indicator at the ego vehicle, $t = -0.89\ s$) to avoid the collision.

Before any assistance functions are executed, a right-of-way sign appears above the relevant vehicle ($t = -2.88\ s$). It indicates the right-of-way relation to the ego vehicle. This relation is the query result after reasoning of the knowledge-base and forms the basis for intersection DAS functions.

The simulation was performed in real-time, showing the capability of the framework to use a traffic situation knowledge-base for in-vehicle driver assistance[43].

5.2.4.3 Incorporated Knowledge-Base Information

Fig. 45 shows an ABox graph excerpt containing its classified objects and relations. It illustrates the knowledge contained in the ontology during the approach to the intersection as depicted in **Fig. 44** (omitting attribute assertions for clarity). The ego vehicle is the starting node of the graph. As well as car 6, it is assigned the concept Car[44]. Originating from the ego vehicle, there is a variety of relations ending at other instances such as a road, lanes, the intersec-

---

[43] Note, that the vehicle simulation environment enables setting a factor of simulation time, so that for larger ontologies and respective reasoning times the simulation could be run slower and real-time capability could be pretended. For example, assuming a reasoning and answering time of 1s would mean, running the simulation with a time factor of 0.5 would assume a reasoning and answering time of only 0.5s. However, all TISDO simulations presented here were run in real-time, that is a time factor of 1.0.

[44] Note, that the numbering of car 6 is exemplary and may be assigned an arbitrary number. The number results from an internal counter, enumerating detected cars during the simulation run.

tion and car 6. Similar relations apply for car 6. The fully reasoned ABox contains a yield relation and a conflicting relation from the ego vehicle to car 6 that is of special interest for driver assistance functions.

In contrast to the variety of relations within the fully reasoned ABox, only a few relations have to be sent to the ontology. In advance, some concept assertions have to be made, i.e. road instances have to be of type Road, lanes of type LaneEntering, LaneExiting or LaneTwoWay, vehicles of type Car and traffic signs of type RightOfWaySign or YieldSign. The sent relations are isOn-relations from vehicles to their lanes, isPart-relations from lanes to roads, connectedTo-relations from roads to the intersection, angle attributes of roads and lanes at the intersection, isPart-relations from traffic signs to their roads and the intersection and approachesTo- or departsFrom-relations from the vehicles to the intersection.

5.3 Knowledge-Based Driver Assistance Functions

5.3.1 Application Scenarios and Functions

Currently, the ontology is optimized for intersection assistance. However, it is easily extendable to other applications where knowledge about the situation is required.

With the above-introduced ontology, DAS functions in addition to existing functions that can be realized or enhanced and situations that can be addressed include:

- Handling of a drivers' or other cars' disregard of right-of-way
- Support in case of an impossible comfortable de-escalation of an upcoming collision at an intersection or on a lane, including collision warning, avoidance braking or evasion
- Information about traffic light phases
- Disregard of red light / stop sign
- Speed limit assistance, curve speed assistance
- Lane departure warning and lane keeping assistance
- Blind spot warning and lane change assistance
- Assistance for stationary obstacles with lane association

These listed points are examples, in which the ontology may have direct utility to be used in the modeled DAS functions. Specific queries may be asked about the situations and, if certain answers return, functions may be executed. Therefore, these functions are expected to be primarily modeled in the form of expert rules, like "if ... then ..." constructs.

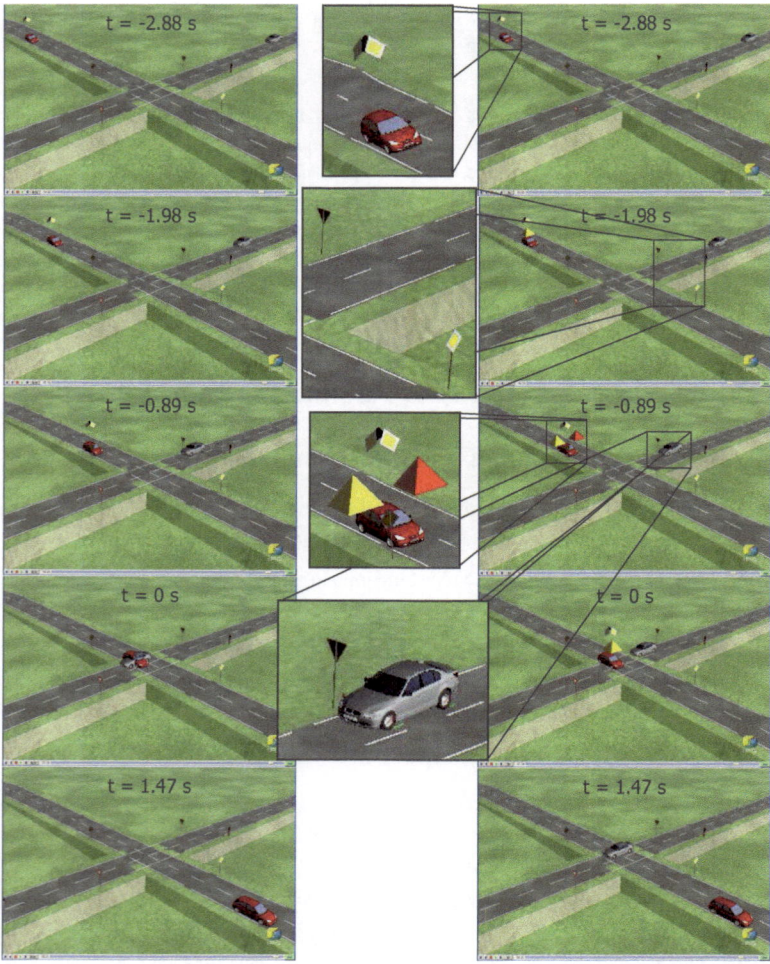

Fig. 44: Comparison of IDAF real-time simulation results for knowledge-based intersection DAS functions. The left side shows simulation with deactivated, the right side with activated DAS functions. At first (t = -2.88 *s*), a right-of-way sign above the approaching red vehicle (ego vehicle silver) shows reasoned right-of-way information from the ontology (see traffic signs in enlargement). Subsequently (t = -1.98 *s*), the yellow pyramid above the red vehicle indicates that the ego vehicle receives an uncomfortable collision avoidance warning, concerning the red vehicle. Finally (t = -0.89 *s*), the red pyramid indicates an automatic full braking of the ego vehicle to avoid a collision (t = 0 *s*). After de-escalation (t = 1.47 *s*) the vehicle starts up again. Both vehicles are initially travelling at a speed of 50 km/h.

Fig. 45: Excerpt of the fully reasoned knowledge (ABox graph) contained in the RacerPro ABox of the example simulation in **Fig. 44** at $t = -1.98\,s$, when enough information to reason right-of-way of car 6 (red car in simulation) is given. It shows that the egocar has to yield car 6 (hasToYield-relation) and that it is conflicting it from left. Also the information about traffic signs is incorporated, yet only partly visible in this graph excerpt but, however, responsible for the right-of-way status of car 6.

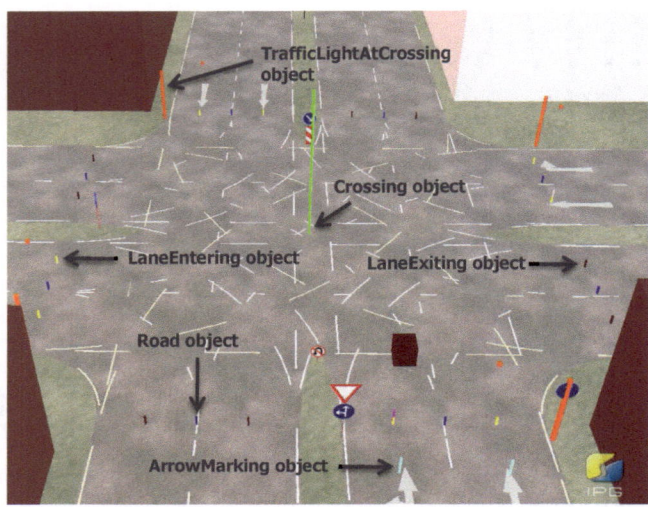

Fig. 46: Representation of relevant intersection objects that can be detected by simulated sensors in the CarMaker simulation environment. Once detected, they and their attributes are transformed into DL statements and sent to the reasoners ABox by the IDAF module.

5.3.2 Exemplary IDAF-Realizations of TISDO-based Driver Assistance Systems

The aim of the IDAF module is to facilitate real-time application of the traffic intersection situation description ontology (TISDO). TISDO is capable of describing traffic situations at complex intersections, as explained in section 5.1. To provide a proof-by-implementation, the IDAF implementation was extended to more sophisticated DAS functions by two associated theses [Hesser, 2011, Spinda, 2011] in two categories: traffic light assistance and turn prediction for enhanced collision avoidance and turning support. The following subsections explain both systems.

5.3.2.1 Simulation and Situation Setup

Part of a situation setup in the simulation environment is the preparation of intersections with various objects to be detected. IDAF must be able to detect an intersection with associated roads, lanes (entering and exiting), traffic signs and lights or arrow markings and even moving traffic participants or other obstacles. Virtual objects, invisible during runtime, according to the mentioned objects were created for this purpose. This is illustrated in **Fig. 46** for a medium size, complex intersection. These objects, illustrated by colored boxes of different size[45], are associated with type-specific attributes, such as object type, position, velocity, cur-

---

[45] Size and color of the virtual objects do not matter during simulation and solely serve to distinguish between types for modelers and observers.

rent state, e.g. the current and future traffic light states, and so on. Buildings are modeled as well, but ignored by SDF after perception by sensors in these example applications. They provide no relevant information to the ontology. However, they would be interesting and could be handled, when occlusion would be considered.

The virtual objects are necessary, because vehicle environmental sensors in the simulation environment do not directly "see" objects such as an intersection, roads, and lanes and so on. The virtual objects, however, can be recognized and, within the ontology, are then semantically identical to the real objects. In vehicle application these objects would be detected by advanced environmental sensors, such as imaging, radar or lidar sensors. These sensors, yet, are still under research and development to provide enough information.

5.3.2.2 TISDO-based Traffic Light Assistance

The first implemented example application is an advanced traffic light assistance, based on abstract information and logical inference within the ontology. In addition, more frequently updated vehicle dynamic data is used for vehicle dynamics prediction and decision making on dynamic actions (compare IDAF module in **Fig. 40**). The abstract ontological information is provided after reasoning is finished, whereas vehicle dynamic data is of latest up-to-date kind, directly forwarded from SDF.

Implementation was part of the associated master thesis [Hesser, 2011]. Scenarios are simulated at a standard four-road intersection with 8 lanes. Dynamic data only contains position, velocity and acceleration of traffic participants, relative to the intersection. Other information, such as the mentioned objects with type, position and state, is stored in the ontology and used for logic reasoning.

The developed traffic light assistance covers several sub-functions:

- Ego traffic light assistance (illustrated in **Fig. 47 a)**): This assistance comes into action, if the traffic light associated to the ego vehicle is currently red, or – if the future state is known – if the ego traffic light will be red, when the ego vehicle will pass the traffic light. In case of a necessary uncomfortable braking maneuver, a warning is executed and, when it becomes necessary, an autonomous emergency braking maneuver is triggered (illustrated by the red box above the vehicle).

- Comfort traffic light assistance: In case of a known future traffic light phase, a recommendation for a comfortable stop can be calculated and provided for the driver. This information could be extended to an economic stop (e.g. as in [Schuricht et al., 2011]). In the case that a vehicle stop is not necessary, a speed recommendation can as well be provided. The speed may lie in a certain interval, e.g. between about 2/3 of and the full maximum allowed speed[46]. Following the recommendation guarantees to hit the green phase.

---

[46] This is just an example but may be recommendable, so that overall traffic is not interfered with by driving too slow.

- Assistance, if another car disobeys its own red light phase and a collision becomes probable (illustrated in **Fig. 47 b)**): In this case, a warning and a speed adaption may be triggered to avoid a collision in a similar manner as with the ego traffic light assistance.

- Considering following cars for avoidance action: For example, if a following car is too close to perform an avoidance braking maneuver, an avoidance acceleration maneuver is calculated and checked. This, obviously, is too dangerous and less applicable for plain assistance, but interesting for autonomous avoidance actions or autonomous vehicle takeover.

All four listed functions were successfully implemented and tested in various scenarios, running in real-time.

5.3.2.3 TISDO-based Turn Prediction and Improved Collision Assistance

The second implemented example application is a turn prediction for enhanced collision avoidance and turning support. Analog the first application, abstract information about the intersection situation generated in the SDF component is stored in the ontology and queried after logical inference. Vehicle dynamic data is used in parallel to dynamically predict and react on situations (compare IDAF module in **Fig. 40**). Implementation of this application was part of the bachelor thesis [Spinda, 2011].

A main functionality of this approach is to provide a turn prediction based on a generic intersection description. During the approach of the ego vehicle towards an intersection, a query about all possible exit lanes is posed. Based on the query results, calculations of possible driving trajectories are carried out. With these trajectories, typical turning speeds can be calculated for the exit lanes.

Here, for ease of calculation, trajectories are modeled by straight–circle segment–straight line paths with the largest possible radiuses between the entering point (straight line tangential) and each exiting point (tangential). More complex alternatives, for example, would be using a clothoidal-based path or using empirically determined trajectories with underlying probabilities.

For the applied radius-based approach, typical turning speeds can be determined, parametrized by common lateral accelerations for lethargic, normal and sportive drivers. With these parametric speeds, intervals for an expected turning speed for each exit lane can be created (non-overlapping). Alternatively, probabilities for each exit lane depending on the turning speed could be derived when using a probabilistic turning speed model, e.g. from empirical data.

In addition to calculating turning speeds, the speed on arrival at the intersection is predicted by using current dynamic data and longitudinal dynamic formulae. For better accuracy, the change of longitudinal acceleration / deceleration is extrapolated linearly. This extrapolation improved accurate prediction of a left turn in the simulation from $1.3\ s$ before entering the intersection to $3.1\ s$ before. The predicted speed at arrival at the intersection is directly used to predict the expected turn using the calculated speed interval for each possible turn.

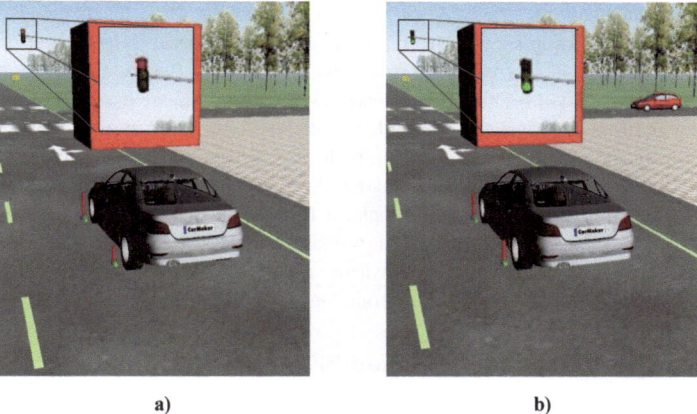

a)   b)

Fig. 47: a) Snapshot of an emergency braking action (vehicle dynamic bars at wheels) and warning (indicated by red box) by the knowledge-based traffic light driver assistance system, when the own red traffic light is disobeyed. **b)** Snapshot of a similar action of the same system, when another vehicle crossing the own vehicles driving path is disobeying its associated red traffic light.
(Traffic lights are enlarged in an overlay for better visibility in this figure.)

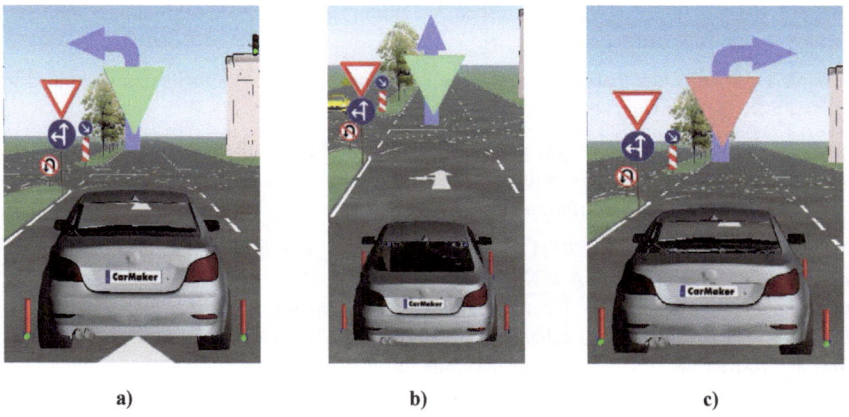

a)   b)   c)

Fig. 48: a) and **b)** Prediction of an upcoming left turn / driving straight maneuver based on the semantic information retrieved from the situation knowledge-base in combination with dynamic vehicle data. The green triangle indicates an allowed maneuver. **c)** Accordingly, prediction of an upcoming right turn maneuver. The red triangle indicates a prohibited maneuver.

At first, ontological information is used to determine the connected lanes to the intersection and their location for turn prediction. Then ontological information is used to inform about legitimacy of the expected turn as depicted in **Fig. 48**.

With information about the expected exit lane that will be used, the collision avoidance introduced in section 5.2.3 can be extensively enhanced. It can be used at large intersections, as such depicted in **Fig. 46**. Moreover, cases can be distinguished, where a collision is at high risk and where a collision is unlikely to happen, even though cars are passing closely. An according comparison is provided in **Fig. 49**. **Fig. 49 a)** shows a left turn of the ego vehicle leading to a collision, whereas in **Fig. 49 b)**, both cars are simply passing each other. The ontological information is necessary to evaluate the vehicles' options with respect to the infrastructure and traffic legislation. For example, a few seconds earlier in **Fig. 49 a)**, nobody would expect the vehicle to cross the green median strip, because there is no lane. A left turn is only possible and potentially expected, where crossing an intersection is actually possible. All this information is contained within the ontology.

5.3.3 Benefit of Knowledge-Based Situation Description

Besides logic reasoning about possible path conflicts, which may result in a potential collision, as well as right-of-way dependencies between vehicles, the ontology TISDO especially provides the distinction, if assistance is necessary at all. This particularly affects assistance using dynamic driving data. An illustration is given in **Fig. 50**. When considering only dynamic driving data of both vehicles, as maybe delivered by some sensors, the situation looks exactly the same. Incorporating knowledge about intersections, roads, lanes and according relations, the situations can be distinguished and appropriate assistance is possible, simultaneously reducing false alarms.

Possible DAS functions that may be addressed with the ontological information contained in TISDO were mentioned in section 5.3.1.

However, the queried information may also serve to build up subsequent models for more complex interpretations or functions, which represent more generic approaches or computation relies on more complex information.

For example, [Brechtel et al., 2011] propose a generic approach for "high-level decision making in traffic environments". A POMDP (Partially Observable Markov Decision Process) is used to plan an optimal policy for maneuvering actions. However, the environment has to be discretized for calculation and calculation is time consuming. To find appropriate discretizations and restrict the search and state space, the retrieval of ontological information about the situation and the environment may be useful.

The same proposal of ontological information support may apply for spatial probability distribution prediction or trajectory prediction of traffic participants. Existing prediction methods are, for example, "Stochastic Reachable Sets of Interacting Traffic Participants" by [Althoff et al., 2009], "Monte Carlo Road Safety Reasoning" by [Broadhurst et al., 2005] and "Accident Prediction Using 3-D Model-Based Vehicle Tracking" by [Hu et al., 2004], which makes use of multi-layer Neural Networks to learn trajectories.

a)   **b)**

Fig. 49: **a)** Collision avoidance emergency warning (yellow) and braking (red) action performed by the knowledge-based enhanced collision and turn driver assistance system when a turn is predicted (case **Fig. 48 a)**). **b)** Omitted collision avoidance action by the same system when no turn is predicted (case **Fig. 48 b)**).

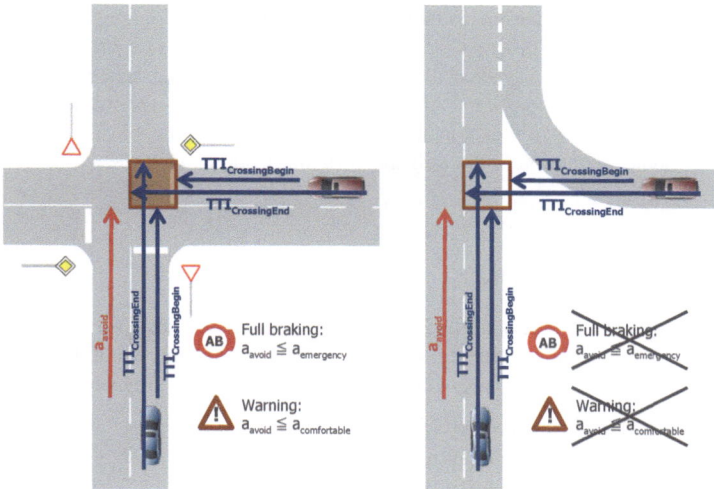

Fig. 50: Comparison of similar cases by dynamic data – one leading to a collision at an intersection, the other resulting in two cars driving next to each other. These situations can be distinguished with the use of knowledge-base information.

The methods discussed in the previous two paragraphs may benefit from extending their models to incorporate ontological information. In contrast, probabilistic methods based on first order logic (FOPL), necessarily need information formulated in first order logic sentences. Typically w.r.t. ADAS, these sentences would contain ontological information. Recently, [Nienhüser et al., 2011] and [Schamm and Zöllner, 2011] applied MLNs and OPRMs to situation interpretation. These methods directly used situative context as input. This situative context may be provided by a situation description ontology, such as TISDO, in the most practical way.

5.4 Uncertainty Handling

Besides being able to describe a traffic situation with prior known and certain information, which was described in the previous sections of this chapter, handling uncertainty is a key issue to cope with. The presence or absence of an object or a relation, for example, may completely change the semantics of a situation. Imagine an intersection with or without vehicles, with or without road signs, with or without multiple lanes or even with or without certain roads. A lack of information or uncertainty about information makes understanding the traffic situation much more demanding.

Section 3.2.5 discussed several methods combining probabilistics and logics to simultaneously cope with semantic knowledge and probabilistic uncertainty. It was pointed out that, unfortunately, recent methods are not yet capable to be used for a generic situation description.

As a consequence, this section will show an approach to deal with some types of uncertainty of sensor information when using a knowledge-based situation description approach. At first, subsection 5.4.1 describes different types of uncertainty and provides some probabilistic formulations for semantic information of ontologies. Subsection 5.4.2 briefly discusses some issues of uncertainty that may come along with axioms and rules. The main part of how to handle prior probabilities of sensor inputs is covered by subsection 5.4.3. The last subsection 5.4.4 adds another approach to cope with unknown information

5.4.1 Sensor Uncertainty

5.4.1.1 Types of Uncertainty

A good overview about uncertainty and imperfect information is researched by and comprised in [Fuchs, 2008]. Therein, uncertainty is defined as "lacking certainty about the truth of some fact". Moreover, more detailed types of uncertainty and imperfect information are given with:

- *Vagueness* as not being well defined,
- *Inconsistency* as having contradictory pieces of information,
- *Ignorance* as lacking information,
- *Imprecision* as failing to be adequately specific and
- *Ambiguity* as not being able to distinguish between two alternatives.

When using ontologies for knowledge-based situation description, these types of uncertainty can be addressed to different extents.

Vagueness can be obviated with thorough and comprehensive ontology engineering. All information of interest in traffic situations that is necessary for subsequent applications and operations should be described in the TBox and further detailed and constrained with axioms. Vagueness that resulted during ontology engineering, however, may be hard to uncover in ontologies, due to the OWA that allows for information not to be known. Thus, even modeled axioms may not be applied, if information is vague and no notification will be given. Developed ontologies should be carefully checked for missing information or unintended taxonomies or ABox results.

Ontologies are excellent in discovering *inconsistencies*, both for TBoxes and ABoxes, to the extent that axioms are given to describe semantics and to set constraints. The reasoner will discover any logical contradiction within the TBox itself or in the ABox with respect to the TBox (see inference services in section 3.2.2). Note that this also accounts for the reasoning results after rule execution, but not for ABox augmentation rules themselves.

The same problem as with vagueness occurs with *ignorance*, which may be a lack of information in the TBox such as subsumption axioms, closure axioms or constraints. In this case, inconsistencies may not be discovered or certain information may not be retrieved from the ontology. Ignorance may also be a lack of situative knowledge within the ABox, i.e. lacking sensory information. Due to the OWA this lack of information will not be discovered as the information is assumed to be possibly existent. This lack of information may be locally discovered and handled with non-monotonic reasoning by rules, e.g. using negation as failure (see sections 3.2.4.3 and 5.1.2.5) or implementing the – computationally more expensive – local closed world assumption LCWA as introduced by [Hummel, 2009].

Imprecision and *ambiguity* both may occur with ontology elements such as objects or relations or with attributes of ontology elements. Attributes may be modeled with a probabilistic notation, i.e. probability distributions or histograms etc. or with a fuzzy notation both offering a variety of well-known algorithms for further treatment.

Subsumption elegantly allows to deal with some forms of imprecision and ambiguity of ontology elements. It may be sufficient to classify an object as the parent concept (or relation as the parent role), instead of having uncertainty in the classification at the child level. For example, a moving traffic object might be known to be a vehicle, but with the exact type, e.g. car, truck, bus etc., still being unknown.

Handling imprecision and ambiguity of objects and relations with probabilistic logic notations is still under research (see section 3.2.5). All of the known methods are highly computationally expensive, because they internally span a large space of possible worlds for probabilistic reasoning.

A more simple and thus more efficient, yet probabilistic approach that considers the characteristics of uncertain sensor information will be discussed with this thesis in section 5.4.3.

A different approach to cope with some cases of ignorance or ambiguity, using additional ABox information to pretend the existence of a real object, will be provided in section 5.4.4.

5.4.1.2 Probabilities of Uncertain Ontological Sensor Information

The different types of sensor information are named in section 2.1.2. They comprise attribute features, object features, classification features and more sophisticated situation features (including further classification as well as relations).

Probabilistic formulations are meaningful for sensor measurement values stored as attributes in the ontology, for existence probabilities of objects and relations as well as for classification probabilities of those accounting for the uncertainty types imprecision and ambiguity.

Attributes may be provided with either continuous or discrete probability distributions. Continuous attributes, e.g. the measured value of an objects velocity, may be provided with continuous probability distributions such as Gaussian mixtures, which may be gained from tracking algorithms such as a Kalman filter, empirically determined distributions or other assumptions. Discrete attribute probability distributions, e.g. for weather conditions, object properties, different discrete system states, etc. could, for example, be obtained from empirical data, fuzzy divisions of signals or other assumptions.

Probabilistic Notations

The remainder of this subsection introduces formalizations of probabilistic notations for semantic elements to be further used for probabilistic uncertainty handling of ontologies in the remainder of this and the next sections. Probabilities are denoted by P, probability distributions by p.

The probabilistic formalization of the probability distribution p for an attribute $a{:}\,A$, provided the underlying object or relation is existent, is given by

$$a{:}\,A,\ (x,a){:}\,R_a,\ x{:}\,C, R :\ \ p\bigl(a|(x,a)\bigr)\ . \tag{54}$$

The existence probability for an unclassified object $c{:}\,\top$ can be denoted with

$$c{:}\,\top :\ \ P(\exists c{:}\,\top) = 1 - P(\nexists c{:}\,\top)\ . \tag{55}$$

This existence probability of a detected object (or relation) may vary over time, e.g. due to tracking quality measures and occlusion, weather conditions etc. Because logic reasoning in TISDO is carried out for a snapshot of a situation, that is a certain point of time, temporal notation and handling is not necessary in this case.

Similarly to the existence probability, the probability for an object classification $c{:}\,C \sqsubseteq \top$ can be given:

$$c{:}\,\top :\ \ P(c{:}\,C) = 1 - P(c{:}\,\neg C)\ . \tag{56}$$

The classification probabilities, i.e. probabilities for the class of the object it belongs to, have to be provided by sensor fusion and processing.

In fact, the existence probability can be interpreted as a classification of c belonging to the top concept $c{:}\,\top$ for existence or as belonging to the bottom concept $c{:}\,\bot$ for non-existence and it can be rewritten as a classification probability:

$$P(\exists c{:}\,\top){:}= P(c{:}\,\top) = 1 - P(c{:}\,\neg\top) = 1 - P(c{:}\,\bot)\ . \tag{57}$$

Comprising Probabilities

Note that classification probabilities cannot be added and do not sum up to 1 by default. This is due to concept subsumption (hierarchy), OWA and allowed object membership of multiple concepts.

The conditional classification probabilities only sum up to 1 by definition for *disjoint concepts* $C_s \sqsubseteq C$ with $\forall i \neq j: C_i \sqsubseteq \neg C_j$ included in a *closure axiom* $C \equiv C_1 \sqcup ... \sqcup C_n$, i.e. these concepts do not overlap and only these concepts belong to their parent concept, provided the object belongs to the parent concept:

$$\sum_{\substack{s, C_s \in \{C_1,...,C_n\}, \\ C \equiv C_1 \sqcup ... \sqcup C_n, \\ \forall i \neq j: \ C_i \sqsubseteq \neg C_j, \ C_i, C_j \in \{C_1,...,C_n\}}} P(c:C_s | c:C) = 1 \ . \tag{58}$$

Without the closure axiom the equality ($= 1$) would become an inequality (≤ 1).

(58) can be rewritten (multiplying with $P(c:C)$), keeping the disjointness and closure constraint, as

$$\sum_{\substack{s, C_s \in \{C_1,...,C_n\}, \\ C \equiv C_1 \sqcup ... \sqcup C_n, \\ \forall i \neq j: \ C_i \sqsubseteq \neg C_j, \ C_i, C_j \in \{C_1,...,C_n\}}} P(c:C_s) = P(c:C) \ . \tag{59}$$

Otherwise, due to the OWA, objects of other concepts may belong to the parent concept and objects may belong to several concepts, unless explicitly stated.

For example, the concept Road may contain the child concepts Highway, RuralRoad, UrbanRoad and ConstructionRoad (see illustrated in **Fig. 51 a)** at first omitting Bridge, PlainRoad and Tunnel for ease of discussion). Yet, only the first three concepts are disjoint, all of them may also currently be construction roads. Hence, the probabilities of the first three child concepts must sum up to 1, if and only if an object is known to be of type road and an axiom states that construction roads have to belong to one of the first three as well. It is recommended to introduce further mid-layer concepts, so that as many closure axioms as possible can be stated.

For the general case of possibly *overlapping concepts* $C_s \sqsubseteq C$ (non-disjoint), the probabilities $P(c:C_s)$ have to be handled the same way as in a probabilistic random experiment.

The probability $P(c:C)$ is equal to the probability for the existence of at least one classification $c:C_s$ and can then be calculated with

$$P(c:C) = 1 - \prod_{s, C_s \sqsubseteq C} \left(1 - P(c:C_s)\right) \ . \tag{60}$$

Note, that the sub-concepts must not be mixed in terms of disjointsness or non-disjointness. Otherwise the result contains the assumption that the included disjoint concepts could overlap.

In these cases of mixed disjointness and non-disjointness, another abstract intermediate-level concept should be introduced that contains all disjoint concepts. See for example **Fig. 51 a)**, where the concept Road subsumes all of Highway, RuralRoad, UrbanRoad, ConstructionRoad and Bridge, PlainRoad and Tunnel and the first three and the latter three concepts each are disjoint. Two abstract intermediate-level concepts RegionalRoadType and RoadBuildingType containing the first and the latter three disjoint sub-concepts may be introduced as illustrated in **Fig. 51**

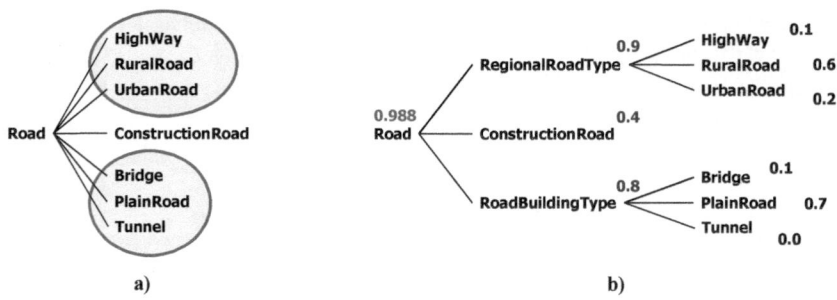

Fig. 51: Example taxonomy for concept Road: **a)** without intermediate-level concepts. Circled concepts are made disjoint among each other; **b)** with inserted intermediate-level concepts for proper classification probability calculation. Black numbers are prior classification probabilities (e.g. provided by a sensor), grey numbers are summed posterior classification probabilities (calculated with (59)). The orange classification probability is calculated with (60), but its correctness depends on axioms about the three sub-concepts.

b) including probability examples. (59) then applies to the concepts Highway (0.1), RuralRoad (0.6) and UrbanRoad (0.2) to sum up the probability for RegionalRoadType (0.9) as well as the latter three types for RoadBuildingType (0.8). Road then subsumes RegionalRoadType, RoadBuildingType and ConstructionRoad. (60) now applies to all of these three non-disjoint concepts to gain the probability for the abstract concept Road (0.988).

However, if an (appropriate) axiom would state that ConstructionRoad implies, its instances also have to belong to RegionalRoadType and RoadBuildingType, (60) cannot be applied. In this case the three concepts are not conditionally independent. This case would rather be handled by the reasoner itself. The reasoner would then have to be extended to inherent probabilistic reasoning, if the implication is not known in advance and explicitly used to gather probabilities and calculate the probability for the classification of the concept Road.

The formulae to comprise probabilities like (59), (60) are provided to restrict the number of possible cases that may have to be regarded. Handling such possible cases is the subject of section 5.4.3. In this way, it has to be carefully checked, where and if such probabilities may be comprised for simplification of further reasoning. Otherwise, distinct cases have to be treated and reasoned separately.

Despite some errors in calculation, other and more specific probabilistic cases, handling different sources and classification tasks especially in the context of sensor fusion, are provided by [Fuchs, 2008]. Depending on the type of input data, the probabilistic handling of ABoxes introduced in the subsequent sections may be extended by the formulae provided in [Fuchs, 2008]. It also includes the belief theory by Dempster-Shafer, an approach for sensor fusion based on probabilistic theory that considers both degree of belief and plausibility for a statement.

5.4.2 Certainty and Uncertainty of Ontology Axioms and Rules

The traffic situation description as delimited in section 2.3 and hence TISDO introduced in Chapter 4 of this thesis only contain well-defined, definite and deterministic reasoning. As such, no uncertain axioms or rules are included by definition. This does not require any probabilistic handling of axioms or rules. Probabilities are only assigned on ontology input (sensor information in general), not during reasoning, which enables the use of efficient existing non-probabilistic reasoners.

In contrast, the traffic situation interpretation may contain uncertain axioms or rules (see section 2.3 and the discussion of probabilistic logic reasoning in section 3.2.5). This may be necessary e.g. for reasoning about behavior, future situation progress and future decisions, where uncertain events may occur, and the mindset of drivers and other traffic participants are not clear.

As well, the part of SDF creating semantic information (classification, relation assignment) has to cope with uncertainty, e.g. when assigning vehicles to lanes / roads or assigning traffic signs to their according lanes or determine which traffic sign is currently relevant based on time, weather condition, position etc. (compare [Nienhüser et al., 2011] for relevance estimation and assignment of traffic signs). Some probabilistic formulations for semantic sensor inputs w. r. t. ontologies were shown in the previous subsection 5.4.1. If these formulations are sufficient to define distinct cases assigned with probabilities, the deterministic TISDO may be used for each case for a simplified probabilistic handling of sensor inputs. This will be discussed in the next subsection 5.4.3. If this is not the case, more sophisticated methods including probabilistic logics have to be applied. These methods are, however, not yet widely applicable, which was subject to the discussion in section 3.2.5.

5.4.3 Handling Prior Probabilities with Ontologies

The knowledge-based traffic situation description approach in this thesis only considers axioms and rules with a deterministic outcome within the ontology, so that for given input information sent to the ABox there is no ambiguity about the resulting fully reasoned ABox. However, as described previously in section 5.4.1, sensor information may come with uncertainty. Because sensor information serves as input to the ontology, a method to handle these probabilistic input formulations will be discussed in this section. This may simplify the computationally expensive probabilistic ontology reasoning.

5.4.3.1 General Case with Classification and Attribute Probabilities

The basic approach is to create different ABoxes for all possible manifestations of uncertain inputs. Because these different manifestations occur jointly, all combinations of different manifestations have to be taken into account. It is inspired by the approaches of Probabilistic Logic by [Nilsson, 1986] and the extension First Order Probabilistic Logic by [Jaumard et al.,

2006], which consider *possible worlds* based on propositional or first order sentences[47]. The approaches first determine the consistent set of these possible worlds to infer their probabilities (inconsistent worlds must have probability of zero). The following subsections will formalize and further detail this approach.

The goal is to deliver probabilities with the information that can be retrieved from the ontology. The resulting ontology will, however, consist of a number of ABoxes representing the possible worlds or combinations of different manifestations of inputs, respectively. Each query posed to the ontology will then contain all different answers of its possible worlds and come along with one or several probabilities that can be derived from the probabilities of occurrence of the according possible worlds.

Creation of Possible Worlds with Ontology Element Classification Probabilities

Let $\mathcal{A}_{prior}$ be an ABox containing the information sent to the reasoner before reasoning itself is performed. The fully reasoned ABox is simply named $\mathcal{A}$. The ABox $\mathcal{A}_{prior}$ consists of assertions such as

$$\mathcal{A}_{prior} = \{c_i : \{C_s\},$$
$$r_j = (c_m, c_n) : R_t, \ c_m, c_n \in \{c_i\}, \tag{61}$$
$$a_k : A_u, \ q_k = (x_o, a_k) : R_a, \ x_o \in \{c_i, r_j\} \}$$

with objects c_i classified as belonging to a set of concepts $\{C_s\}$, relations r_j between existing objects c_m, c_n of type R_t and attributes a_k of type A_u with a value and a unit. These attributes a_k are part of attribute relations q_k of type R_a originating from objects or relations x_o.

Because the traffic situation description as proposed in this thesis does not contain rules and axioms with uncertainty, but solely uncertainty about sensor information, all elements with uncertain information are already contained in the ABox $\mathcal{A}_{prior}$. This means, there is a subset of objects c_u with $\mathcal{A}_{prior} \vDash c_u : \top$ as well as a subset of relations r_v in the same way[48], i.e. they are entailed by the ABox $\mathcal{A}_{prior}$. These objects and relations contain the sensor information. Thus, both c_u, r_v are provided with classification (or existence[49]) probabilities and, regarding a single element c_u or r_v, a set of ABoxes $\{\mathcal{A}_{prior,u,s}\}$ for objects (or with index v for relations) can be created for each possible classification $c_u : C_s$ that has an according classification probability $P(c_u : C_s) > 0$. The inequality is stated, because reasoning for impossible worlds, where $P(c_u : C_s) = 0$, is *not reasonable*[50].

For simplification, further examinations will be made with objects c_u belonging to concepts C_s, but they refer to both concepts and roles and their elements.

---

[47] Only simple FOL sentences over $\forall, \exists, \neg, \sqcap, \sqcup$ are allowed in those approaches (no implication, equivalence, cardinal restrictions etc.).

[48] Note, that it is $C_s \sqsubseteq \top$, so that $\mathcal{A}_{prior} \vDash c_u : \top$ contains all assertions $c_u : \{C_s\}$ as well.

[49] Existence probabilities are handled as classification probabilities, as explained with (57).

[50] This may be regarded as a pun, meaning either absurd, if logical reasoning is still possible, or impossible, if inconsistency or unsatisfiability would result.

The distinct ABoxes $\mathcal{A}_{prior,u,s}$ and the accordingly resulting distinct ABoxes $\mathcal{A}_{u,s}$ after reasoning (and thus $\mathcal{A}_{u,s} \vDash \mathcal{A}_{prior,u,s}$) will be assigned the probabilities

$$P\left(\mathcal{A}_{prior,u,s}\right) = P\left(\mathcal{A}_{u,s}\right) = P(c_u : C_s) > 0 . \tag{62}$$

Note, that the probabilities of all ABoxes $\mathcal{A}_{prior,u,s}$ and $\mathcal{A}_{u,s}$, respectively, will not necessarily sum up to 1, as explained in 5.4.1.2.

As there are n_u ontology elements c_u assigned with $\mathbf{n}_s(u)$ classification probabilities each[51], an ABox space of possible worlds is spanned. Combining each possible realization in terms of classification of the ontology elements, the number of possible worlds $\mathcal{A}_{prior,s}$ with $\|\mathbf{s}\| = \|\mathbf{n}_s\| = n_u$ is calculated[52] to be

$$n_{\mathcal{A}} = \prod_u \mathbf{n}_s(u) , \tag{63}$$

building an n_u-dimensional joint probability matrix $\mathbf{P}\left(\mathcal{A}_{prior,s}\right)$ with the elements[53]

$$\mathbf{P}\left(\mathcal{A}_{prior,s}\right)_\mathbf{s} = \prod_{u=1...n_u} \prod_{s=1...\mathbf{n}_s(u)} P\left(\mathcal{A}_{prior,u,s}\right) . \tag{64}$$

Reasoning on each of the ABoxes $\mathcal{A}_{prior,s}$ separately then results in a set of fully reasoned ABoxes $\mathcal{A}_\mathbf{s}$ with $\mathbf{P}\left(\mathcal{A}_{prior,s}\right) = \mathbf{P}(\mathcal{A}_\mathbf{s})$. As explained in section 5.1.3 and shown practically in 5.2, DAS agents may query the knowledge-base to receive information about the situation. Considering classification probabilities, the knowledge-base contains all possible worlds with the according ABoxes $\mathcal{A}_\mathbf{s}$ and the query has to be answered for each ABox separately. Each answer is then assigned the probability of the respective ABox.

Note, that the probabilities of the answers cannot be added up in general. Certain conditions have to hold as explained above and shown in (58) and (59) as well as further detailed in 5.4.3.2. Calculation of the overall probability of a queried answer (e.g. hasRightOfWay between certain objects) is dependent of the type of objects c_u. During probability evaluation, it has to be considered, if e.g. (58) or (59) have to be applied on object c_u. Note, that this is harder, if calculation methods have to be mixed. It is recommended to provide sensor input in a way that e.g. application of (58) or (59) is clear and of one type. The overall probability is then calculated using (58) or (59) (or more sophisticated methods accordingly) along each according dimension u of $\mathbf{P}$.

As an example, consider a situation, where a possibly detected object and one road are not sure to exist, as depicted in **Fig. 52**. There are two (uncertain) objects $c_1 = road_1$ and $c_2 = car_1$. The concepts $\{C_s\}$ are $\{Road, Car, \bot\}$, so that the prior probabilities $P(c_u : C_s) > 0$ are given as shown in the figure. Hence, for each of the objects, there are two possible classifications. The probabilities for each object have to sum up to 1 according to (56) or (57), respec-

---

[51] Element u of quantity vector $\mathbf{n}_s$ contains the number of different classifications $c_u : C_s$ and $P(c_u : C_s) > 0$.
[52] Element u of realization vector $\mathbf{s}$ (or concept assignment vector) contains the specific index s of the assigned concept C_s.
[53] $\mathbf{P}\left(\mathcal{A}_{prior,s}\right)_\mathbf{s}$ is one element of the matrix $\mathbf{P}$ with at a specific position $\mathbf{s}$.

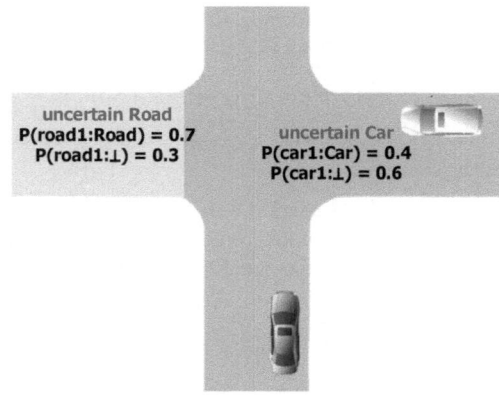

Fig. 52: Simple example intersection with car and one road attached with uncertainty, the remaining elements are assumed to be certain.

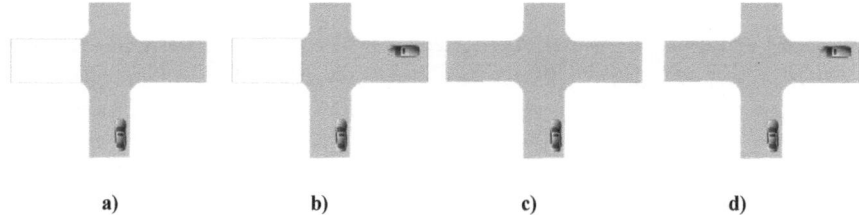

a)   b)   c)   d)

Fig. 53: Possible worlds of the simple intersection example (with calculated probabilities): **a)** $\mathcal{A}_{prior,1,1}$ (0.18), **b)** $\mathcal{A}_{prior,1,2}$ (0.12), **c)** $\mathcal{A}_{prior,2,1}$ (0.42) and **d)** $\mathcal{A}_{prior,2,2}$ (0.28).

tively. With the quantity vector $\mathbf{n}_s = (2,2)$, there are $n_{\mathcal{A}} = 4$ possible worlds $\mathcal{A}_{prior,s}$ with $\mathbf{s} \in \{(1,1),(1,2),(2,1),(2,2)\}$, or distinct ABoxes $\mathcal{A}_{prior,1,1}$, $\mathcal{A}_{prior,1,2}$, $\mathcal{A}_{prior,2,1}$, $\mathcal{A}_{prior,2,2}$, respectively. They are illustrated in **Fig. 53**. With the prior probabilities as given in **Fig. 52**, the probability matrix of the possible worlds is calculated as

$$\mathbf{P}(\mathcal{A}_{prior,s})_s = \begin{pmatrix} 0.3 \cdot 0.6 & 0.3 \cdot 0.4 \\ 0.7 \cdot 0.6 & 0.7 \cdot 0.4 \end{pmatrix} = \begin{pmatrix} 0.18 & 0.12 \\ 0.42 & 0.28 \end{pmatrix}. \tag{65}$$

The association of probability matrix elements is shown in **Fig. 53**.

If a DAS function now queries the ontology with the question, what cars have a hasRightOf-Way relation to the ego vehicle (blue car), ABoxes $\mathcal{A}_{prior,1,1}$ and $\mathcal{A}_{prior,2,1}$ will answer none, whereas ABoxes $\mathcal{A}_{prior,1,2}$ and $\mathcal{A}_{prior,2,2}$ will answer car 1. Because classification probabilities can simply be added in this case, the overall probabilistic answer will be: {(none, 0.6), (car 1, 0.4)}. In this case, these probabilities equal the according probabilities of existence or non-existence of the car, respectively, because the answer only depends on the car and not on the road.

This is not the general case. Consider adding a third car on the left (uncertain) road. Then, the result would highly depend on the existence of the uncertain road itself. A world with the existence of the car coming from left and non-existence of the road must not exist due to the axiom that a car is on exactly one road. Because the road does not exist in this world, the corresponding ABox turns out inconsistent. This ABox will still have a formal probability, resulting from prior probabilities. Because it must not exist, the other probabilities additionally have to be normalized by the remaining probability of consistent words: $1 - P(\text{inconst. worlds})$, i.e. conditioned by the probability of consistent worlds.

Creation of Possible Worlds with Probability Distributions of Ontology Attributes

For attributes with uncertainty or probability distributions, respectively, some aspects are to be considered. If the attributes concerned are not part of any axioms within the TBox they will not affect reasoning or consistency checking and hence do not have to be further treated.

Yet, special treatment is required, if reasoning is based on any attributes and not plainly on logic combination and classification of existing objects and relations and these attributes involve uncertainty. In these cases, the probability for possible concept / role assertions or resulting elements in the consequence of an axiom or rule has to be calculated with integration over the probability distribution of the attribute and an existence probability has to be assigned.

The problem of attributes with probability distributions, especially continuous probability distributions is the prior ignorance of decision boundaries unless the TBox is known and searched for axioms and rules including attributes. This may become even trickier, when transformations of attributes are involved, for example, such as conversion between different units or mathematical operations. Hence, in contrast to handling classification probabilities of ontology elements as described above, attribute probability distributions can only be handled, if the TBox content is sufficiently known in advance, as for example by the modeler himself. Otherwise, more sophisticated probabilistic ontological reasoning systems have to be used like FOPL (see section 3.2.5).

Assuming that all axioms and rules with attributes in their conditions are known in advance, i.e. the TBox is known, then attribute borders (lower bound u_u, upper bound o_u) for attributes a_u can be found as decision borders of the axioms and rules and, hence, neighboring intervals, e.g. $I_{u,s} = [u_{u,s}, o_{u,s}[$ (type of brackets depending of the decision symbol / operator in the axiom / rule such as $>, <, =$), yielding the same reasoning results can be defined.

With the attribute probability distribution $p(a_u)$ from (54), probabilities for the attribute value to be in each one of the interval $P(a_u \in I_{u,s})$ can be calculated. For reasoning, as many exemplary ABoxes $\mathcal{A}_{prior,u,s}$ as intervals $I_{u,s}$ will be created with an exemplary attribute value $a_{u,s} \in I_{u,s}$ accordingly in each one of the intervals. In this case, for each attribute a_u separately, all probabilities of the ABoxes belonging to one attribute will add up to 1:

$$\sum_s P(a_{u,s} \in I_{u,s}) = 1 \; . \tag{66}$$

Succeeding the creation of these intervals, the probabilistic knowledge-base now including attribute probability distributions is handled as described above for classification probabilities, spanning an ontology space of possible worlds $\mathcal{A}_{prior,s}$.

5.4.3.2 Ways to Reduce the Probabilistic Ontology Space

All resulting ABoxes or possible worlds, respectively, may end up different and, in the general case, all of them have to be fully reasoned. Hence, it is interesting to determine, if there are sets of possible worlds that will not be different, to speed up probabilistic reasoning. The following two subsections will show two according approaches.

The first approach benefits from subsumption inherent to ontologies. The second one shows the benefit, if grouping of objects by so-called key objects is possible.

Grouping ABoxes Using Subsumption Characteristics

In some cases such as the application of right-of-way traffic rules, axioms and rules may have been formulated on a more abstract level than the sensor information is actually stored at. Right-of-way traffic rules, for example, as discussed in section 5.1.2, can be formulated at the level of road connections subsuming both lanes and roads as well as at the level of vehicles subsuming e.g. cars, trucks, busses, motorcycles etc.

If axioms and rules are known to be formulated on a minimum level of abstraction, i.e. they plainly contain concepts C_v (and roles respectively) and do not contain less abstract concepts $C_{v,w}$ with $C_{v,w} \sqsubseteq C_v$ and objects c_u are assigned to $C_{v,w}$ with classification probabilities $P(c_u : C_{v,w})$, then different possible worlds in terms of ABoxes $\mathcal{A}_{prior,u,v,w}$ can be conglomerated into one world $\mathcal{A}_{prior,u,v}$ that has a single probability $P(\mathcal{A}_{prior,u,v}) = P(c_u : C_v)$.

For the case of *disjoint concepts* $C_{v,w}$, the probabilities $P(c_u : C_{v,w})$ can be summed up to get the probability $P(c_u : C_v)$ analogously to (59):

$$P(c_u : C_v) = \sum_{\substack{w,\, C_{v,w} \in \{C_1, \ldots, C_n\}, \\ C_v \equiv C_1 \sqcup \ldots \sqcup C_n, \\ \forall i \neq j:\, C_i \sqsubseteq \neg C_j,\, C_i, C_j \in \{C_1, \ldots, C_n\}}} P(c_u : C_{v,w}) \tag{67}$$

For the general case of possibly *overlapping concepts* $C_{v,w}$ (non-disjoint), the probability $P(c_u : C_v)$ can then be calculated as in (60) with

$$P(c_u : C_v) = 1 - \prod_w \left(1 - P\big(c_u : C_{v,w} \big) \right) . \tag{68}$$

For each object $c_u : C_v$, there were $\mathbf{n}_w(u)$ sub-concepts $C_{v,w}$. Because the number of possible worlds is the product of all classification numbers of the objects given by (63), grouping ABoxes using subsumption reduces the number of possible worlds by the product of factors $\mathbf{n}_w(u)$ for each grouped c_u.

Let us assume the concept Vehicle subsumes the four concepts Car, Truck, Bus and Motorcycle. Grouping the uncertain classifications to one uncertain classification Vehicle reduces the number of possible worlds by factor 4, for one Vehicle object, by factor $4 \cdot 4 = 16$ for two vehicles and so on. If a non-zero non-existence probability remains, i.e. there is a probability for the concept $\bot$ of such an object, only 4 of the 5 concepts can be grouped. Hence, a factor 2 (Vehicle or $\bot$) remains instead of factor 5 (Car, Truck, Bus, Motorcycle or $\bot$) and the number of possible worlds is reduced by factor 5/2, exponentiated by the number of such objects.

Grouping ABoxes Using Key Objects of Specific Concepts / Ontology Parts

Let us consider the situation including an intersection and a sensor that detected one right-of-way (or yield) traffic sign with classification probabilities (and an existence probability, respectively) assigned. Legislation restricts infrastructure to be built up in a way that there are right-of-way traffic signs regulating traffic either at all roads or at none[54]. The same applies for traffic lights[55]. Provided the infrastructure is built up correctly, right-of-way signs can be considered to be existent at all roads, if one right-of-way sign is detected.

Hence, it is possible to reduce the number of possible worlds to be reasoned by grouping elements that are e.g. belonging to a specific concept or to an ontology part (sub-graph). In the example, having detected the traffic sign, belonging to an intersection, a group of traffic signs may be created by the sensor and assigned with the existence probability of the detected sign. Accordingly, as many groups as possible classifications of the detected sign may be created instead.

Such an object that all others of a group and the group itself depend on will subsequently be called *key object*. A group G of elements e_i belonging to concepts or roles can be defined as (the key object may be any of the group elements):

$$G = \left\{ e_1 : C_1, e_2 : C_2, \ldots, e_n : C_n, \ e_{n+1} : R_1, e_{n+2} : R_2, \ldots, e_{n+m} : R_m \right\} . \tag{69}$$

The existence of all group elements is depending on each other: given one group element, the remaining group elements have to exist:

$$\forall j, i : P\left(\exists e_j \in G \mid \exists e_i \in G \right) = 1 \ \wedge \ P\left(\exists e_j \in G \mid \nexists e_i \in G \right) = 0 . \tag{70}$$

Therefore the entire group has the existence / classification probability of each single element itself:

$$\Rightarrow \forall j, i : P\left(\exists e_j \in G \right) = P(\exists e_i \in G) = P(\exists G) = 1 - P(\nexists G) . \tag{71}$$

Reasoning will then be carried out for the two ABoxes containing the group $\mathcal{A}_{prior, \exists G}$ and not containing the group $\mathcal{A}_{prior, \nexists G}$.

For a key object with different possible classifications, as many groups G_s as possible classifications may be created and assigned with probabilities. In this case, all of the ABoxes entailing each group $\mathcal{A}_{prior, G_s}$ and possibly the ABox $\mathcal{A}_{prior, \forall s : \nexists G_s}$ with non-existence of any of the groups have to be reasoned.

---

[54] [BMVBS, 2009] (translated): "Each intersection and confluence that is intended to deviate from the principle "right before left" rule, is to be signposted both positively and negatively [...]."
[55] [FGSV, 2010] (translated): "Positive and negative right-of-way signs are always to be arranged and to be installed at the traffic light post [...]."

All other combinations of object non-existences or classifications not contained in the group(s) will result in inconsistent ABoxes and thus do not have to be considered for reasoning. An example for non-existence is the absence of one traffic sign. An example for classification other than in the group is, if one object supposed to be a traffic sign is assigned to the concept traffic light or an object supposed to be a right-of-way sign is assigned the concept for yield signs.

Another example for a small group would be the detection of a lane and the according generation of a road object or similar for a detected lane marking with a generated lane object and road object. The generated objects would have the same existence probability and could be grouped[56].

For the example of 4 traffic signs (existence, non-existence) at an intersection, which belong to a group, the factor representing $2^4 = 16$ possible worlds without grouping can be reduced to the factor 2. Hence, the overall number of possible worlds in this case is reduced to 2.

Let us revisit the simple intersection example illustrated in **Fig. 30** on page 95. Further, let us assume the ego vehicle and the ego road (South) given[57], but the three roads (East, North, West) having existence uncertainty (factor 2 each). The two vehicles driving on roads North and West may belong to one of Car, Truck, Bus and Motorcycle or not exist at all (each factor 5). Additionally, 4 traffic signs may exist and belong to RightOfWaySign or YieldSign (each factor 3).

In the naïve case, $2^3 \cdot 5^2 \cdot 3^4 = 16{,}200$ possible worlds (ABoxes) would have to be created to perform probabilistic reasoning about right-of-way. Reasoning out these ABoxes (each ~0.5 s) would take roughly 8,000 s.

However, taking into consideration the idea of grouping, vehicles can be grouped (now each factor 2) and traffic signs can be grouped (now the entire group has a factor of 3). In this way, the number of possible worlds is reduced to $2^3 \cdot 2^2 \cdot 3 = 96$. This is a reduction by factor 168.75. Reasoning out the remaining ABoxes would now only take about 48 s. The reduction is enormous and should reduce to reasonable time consumption w. r. t. real-time within a few years. Provided, roads are retrieved from navigation map data and this data is assumed to be certain, only 12 ABoxes would remain with roughly 6 s of reasoning time.

5.4.4 Ghost-Objects for Unknown Information

In some cases, for example at intersections in densely built areas, some objects may be occluded by a building, vegetation, elevation or other elements so that in-vehicle sensory equipment is not able to detect and create objects. If however, the occlusion itself is detected by separate algorithms or – with a more complex formulation of the ontological model – reasoned, one or several virtual objects may be created to infer a possible right-of-way situation.

---

[56] This applies for detection including this kind of dependency. The other way round, i.e. a road is detected by the sensor, does not necessarily mean, there is a lane as well.
[57] Let us also assume, all according relations are certain of type and existence as well, if their objects exist.

To depict this idea, **Fig. 52** can be referred to, assuming the "uncertain car" would not be detectable due to occlusion. To determine the possible effect of the occlusion, this vehicle would then in a particular case be assumed to be existent and the ontology would be reasoned out with the anticipated object.

Adjusted DAS-functions may then be executed, e.g. to solely warn or inform the driver of a possibly critical situation.

5.5 Summary

This chapter has described and shown the possibility to use extensive knowledge representation combined with logic reasoning and characteristics of traffic situations to facilitate comprehensive advanced driver assistance functions and their development.

Knowledge engineering is suitable to create a traffic situation description ontology for complex traffic situations, especially those at intersections. Hence, a Traffic Intersection Situation Description Ontology (TISDO) was built. Logic reasoning is capable of performing the necessary execution of traffic rules and to check consistency of the modeled knowledge and the situation with respect to certain given constraints (e.g. infrastructure rules, traffic rules or other desired constraints).

The target was to build a generic situation description. TISDO is generic in the sense that it allows for arbitrary road networks with arbitrarily large intersections each consisting of a number of roads and lanes, traffic signs and traffic participants that is not limited. Traffic rules can be formulated generically and be fully reasoned out to subsequently provide situative knowledge to driver assistance functions. Various drive assistance functions pose queries to the ontology for the reasoner to provide answers on these queries. In case of a positive answer or a match on the queried situation, respectively, driver assistance functions can perform their actions.

These assistance functions enabled with the introduced ontology include, but are not limited to assistance handling situations with disregard of right-of-way, enhanced support in case of an upcoming collision at an intersection or on a lane, including collision warning, avoidance braking or evasion combined with lane information, traffic light assistance, red-light / stop sign assistance, speed limit assistance, curve speed assistance, lane departure warning and lane keeping assistance, blind spot warning and lane change assistance and assistance for stationary obstacles with lane association.

It is particularly worth mentioning that there is no need for separate knowledge representation systems, but all information for the above mentioned functions can be modeled and is contained within the same ontology – TISDO – in one place. These functions are facilitated with the ontological approach assuming sufficient sensor data being provided.

Reasoning on a complex intersection is performed in roughly one second. Typical 4-road intersections take reasonably less time, so that such an ontology is not far from the ability to be implemented in real systems.

For testing purposes, a real-time framework was developed and linked to a 3D vehicle dynamics simulation environment. Several exemplary driver assistance functions were successfully performed in real-time, representing some of the above mentioned functions and all using one and the same ontology.

Furthermore, this ontology is capable to handle partial knowledge or to some extent unknown information. A whole section in this chapter dealt with handling of uncertainty that is generally difficult. Some handling of uncertainty is discussed and facilitated and can be accelerated by some of the proposed approaches that were discussed.

Chapter 6

Relevance by Mutual Information on Ontology Features

Chapter 1 delineates situation features and a holistic method to create semantics with these features to be necessary for a generic situation description. Mutual information based feature selection as a method to identify relevant situation features with respect to some priory known target function or class is described in section 3.1. Its application on vehicle and environmental sensor measurement data is investigated and applied in Chapter 4. Ontologies are explained in section 3.2 and found to be suitable for a holistic description of traffic situations in Chapter 5.

Such an ontology contains information about a traffic situation, i.e. especially its classified ontology elements with relations and attributes. These elements can be interpreted as situation features as well. In this way, the ontology content has to have relevance with respect to some target application. An ontology may be oversized or undersized. Therefore, it is of interest to find quantitative measures to determine the relevance of single ontology elements, such as concepts and roles as well as attribute types. Knowing their relevance has the potential to enable effective and efficient ontology engineering and to create lean but capable and expressive knowledge-based systems, especially driver assistance systems.

This chapter is divided into three sections. Section 6.1 first formalizes semantic situation features to be usable for mutual information calculations. Section 6.2 then explains different applications of such calculations and section 6.3 presents real world measurements, which were semantically labeled. Based on this labeled data, semantic features were retrieved and an MI ranking of these semantic features is provided and discussed.

6.1 Selecting Important Ontology Features

At first, subsection 6.1.1 will introduce and formalize the basic feature generation for single ontology element types such as concepts, roles and attribute types from single ontology elements. With the formalized, quantitative features, mutual information can be applied in those ways explained in Chapter 4 to obtain relevance metrics. Subsection 6.1.2 will then define and formalize features for combined ontology element types or ontology subgraphs, respectively.

6.1.1 Features for Ontology Elements

Ontology elements that remain persistent for arbitrary traffic situations are the terms contained within the TBox. Thus, features can be derived for TBox elements, such as concepts, roles and attribute types. Compared to measurement features used for mutual information calculations as in Chapter 4, those persistent elements are the quantities, e.g. the quantity for a vehicles velocity. The mutual information is actually calculated with all given values for these quantities, i.e. all concrete realizations of these quantities. For ontologies, these realizations are the actual existing objects, relations and attribute values in a situation. These concrete ontology elements are given with assertional axioms such as c:C, r = (c,d):R with c,d:C, as defined in **Table 2**. Attributes are defined similarly to relations with a domain value as target: a:A with q = (c,a):R_a and c:C or q = (r,a):R_a and r:R.

For feature generation, it is assumed that a specific situation or type of situation, respectively, is characterized not by the individual object, which may be different in every situation and have a different name, but by the object types (concepts), types of relations (roles) between them, their number of occurrences in a situation and their attributes.

Taking this assumption into account, a quantity n_{C_k}, as the number of occurrences of objects, and a quantity n_{R_l}, as the number of occurrences of relations, of the according concept C_k or role R_l in a situation are defined. Note, that each occurrence must only be counted once for each of its most specific classification, but not for their parent concepts. n_{C_k}, n_{R_l} can be formalized as:

$$n_{C_k} = \sum_{\substack{c:\, c:C_k, \\ \nexists C_i:\, c:C_i,\, C_i \subseteq C_k}} 1 \;,\;\; n_{R_l} = \sum_{\substack{(c,d):\, (c,d):R_l, \\ \nexists R_j:\, (c,d):R_j,\, R_j \subseteq R_l}} 1 \;. \tag{72}$$

Accordingly, the number of attributes of a specific type A_m is defined as

$$n_{A_m} = \sum_{a:A_m} 1 \;. \tag{73}$$

6.1.1.1 Ontology Concept and Role Features

Given the quantities for concept occurrences n_{C_k} in each situation, ontology features for concepts are then defined as:

$$F_{C_k} = n_{C_k} + \sum_{C_i \subseteq C_k} n_{C_i} \;. \tag{74}$$

Hence, the feature F_{C_k} comprises the number of occurrences of the specified concept C_k and all its subsumed concepts C_i. This is reasonable, because any object classified as belonging to a child concept will be classified by the reasoner as belonging to the specified concept using subsumption and the taxonomy. Assuming the example of cars and trucks hierarchically belonging to vehicles, the ontology feature vehicle will contain the sum of occurrences of cars and trucks, as well as all vehicles that cannot be classified more detailed. The role features are defined accordingly:

$$F_{R_l} = n_{R_l} + \sum_{R_j \subseteq R_l} n_{R_j} \ . \tag{75}$$

The ontology concept and role features F_{C_k}, F_{R_l} not only take into account, if the specified concepts or roles or their children are represented in a situation at all (i.e. instances exist). They also account for the number of occurrences, distinguishing even situations with different numbers of objects of the same type, e.g. a situation with only one or two or more vehicles involved.

With these features created, all mutual information based methods known from section 3.1 and from Chapter 4 can be applied. The according rankings will then be built with

$$I(F_{C_k}; C), I(F_{R_l}; C) \tag{76}$$

for a target function or class C.

6.1.1.2 Ontology Attribute Features

Generating features for ontology attribute types is more complex. The situation is characterized by attribute values, not by their occurrence. Thus, attribute information has to be preserved as much as possible. Ontology attribute types correspond directly to situation feature quantities which are used in section 3.1 and Chapter 4 to build mutual information feature rankings. In ontologies those quantities are attached to ontology elements using attributes.

Ontology Attribute Features Preserving Attribute Information

Defining an attribute range for each attribute A_m with

$$\mathrm{rg}_{A_m} := \max_{A_m} - \min_{A_m} + 1 \ , \tag{77}$$

that is the range between the minimum and the maximum value of all underlying data of this attribute[58] or alternatively, if provided, the preliminary range of allowed attribute values, an attribute feature can be defined as

$$F_{A_m} = \sum_{a_i : A_m} (a_i - \min_{A_m}) \cdot (\mathrm{rg}_{A_m})^{i-1} \tag{78}$$

The principle is according to number systems, e.g. the dual system where $\mathrm{rg}_2 := 2$ or the decimal system where $\mathrm{rg}_{10} := 10$ and so on, where a_i represents the digit, i the index / position of the digit and rg the range for one digit.

This method to calculate attribute features preserves information about all attributes of all types in a situation. For example, a situation with two velocities of three objects will be treated separately from a situation with three velocities of three objects. This is, even if all veloci-

---

[58] There is an offset of 1 on the range to take all possible digits in account because one is omitted calculating the difference between the maximum and minimum value.

ties where equal in such a situation. No attribute information, neither attribute values nor the number of attribute values is omitted.

However, there are several big disadvantages in mutual information calculation of

$$I(F_{A_m}; C) \tag{79}$$

for relevance rankings using this type of attribute feature:

- Reoccurrences of the same combinations of attribute values will be very rare with more and more attributes of the same type in a situation. This reduces accuracy in statistical calculations as several occurrences in the same feature value histogram interval are needed for a good estimate of the probability distribution. Hence, an enormous amount of available data would be required.

- Mutual information calculation needs the division of the feature value range into a definite number of intervals to estimate the probability distribution. With this approach, the feature would have to be divided into $h^{n_{A_m}}$ intervals compared to h intervals of a single continuous feature, with h as the standard number of chosen intervals. On one hand, this will significantly slow down mutual information calculations. On the other hand, special handling in the implementation would be needed in addition to consider different numbers of intervals for each attribute feature.

These problems take effect especially, when the number of attributes of a certain attribute type is high in a situation. If the number is held low, e.g. by many specific attribute types, mutual information calculation may possibly still be sufficiently accurate. Another way to reduce the number of attributes that are considered is by connecting the attribute type to a concept which will be shown later in section 6.1.2.

Ontology Attribute Features with Loss of Information

To encounter the problems with composed attribute features from (78), other types of attribute features can be generated. However, these features come along with a loss of information. The following features combine several attribute values into one feature value. One option is to sum up attribute values forming the feature

$$F_{A_m,\text{sum}} = \sum_{a_i:A_m} a_i \ . \tag{80}$$

To some extent, this feature is capable of representing the number of attributes and value levels. Yet, it implies ambiguousness concerning several attributes and their values. For example, many attributes with low values and few attributes with high values may result in approximately the same feature value and be counted in the same interval for probability estimation. This will reduce the mutual information. Assuming sufficient amounts of data to correctly calculate mutual information values, it had to be $I(F_{A_m,\text{sum}}; C) < I(F_{A_m}; C)$. However, due to the delineated problems, this will not be the general case.

With the number n_{A_m} of attributes of a specific attribute type A_m from (73), an alternative to the feature with plain attribute summation is calculating their average with

$$F_{A_m,\text{avg}} = \frac{1}{n_{A_m}} \sum_{a_i:A_m} a_i \ . \tag{81}$$

This average feature better accounts for the level of attribute values. Both features do not consider differences in attribute values, so that they make a standard deviation feature useful that is defined as

$$F_{A_m,\text{std}} = \sqrt{\frac{1}{n_{A_m}-1} \sum_{a_i:A_m} (a_i - \bar{a})^2} \ \text{ and } \ \bar{a} = \frac{1}{n_{A_m}} \sum_{a_i:A_m} a_i \ . \tag{82}$$

Rankings are then built with

$$I(F_{A_m,\text{sum}}; C),\ I(F_{A_m,\text{avg}}; C) \text{ and } I(F_{A_m,\text{std}}; C) \ . \tag{83}$$

Because this treats the information of the generated features separately, better information about the relevance of attribute features would be obtained by considering the joint mutual information

$$I(F_{A_m,\text{avg}}, F_{A_m,\text{std}}; C) \text{ or } I(F_{A_m,\text{sum}}, F_{A_m,\text{std}}; C) \tag{84}$$

in the ranking. Because the joint mutual information is not calculated between all pairs of features but feature pairs of each specific attribute type, this does not unmanageably increase ranking times in contrast to the methods proposed in section 0.

As well as using the standard deviation feature, the rankings could include joint mutual information with a plain attribute number feature such as

$$I(F_{A_m,\text{avg}}, F_{n_{A_m}}; C) \text{ or } I(F_{A_m,\text{sum}}, F_{n_{A_m}}; C) \text{ with } F_{n_{A_m}} = n_{A_m}. \tag{85}$$

6.1.2 Combining Ontology Elements as Element Chain Features

The previous section introduced how features describing occurrences of different ontology element types can be derived to determine their relevance for certain applications. Situations, however, are not only specified by single ontology element types, but their semantics is especially formed with combinations of those elements, i.e. certain concepts and their role occurrences among each other. Consequently, it is of special interest to form features containing this kind of information. This subsection will form features to serve this goal.

6.1.2.1 Considering Simple Relations of Ontology Elements

Subsection 6.1.1 defined quantities regarding the numbers of occurrences of single element types as in (72). Similarly, quantities representing the number of combinations of objects of a certain concept with relations of a certain type to other objects can be defined. The same can be done for relations of certain types to other objects of certain types. Quantities $n_{C_k \cdot R_l}$, $n_{R_l \cdot C_p}$, $n_{C_k \cdot R_l \cdot C_p}$ hence represent occurences of combinations of concept C_k with a role R_l, role R_l with concept C_p and concept C_k with role R_l with concept C_p. These quantities are defined by

$$n_{C_k.R_l} = \sum_{\substack{c,d:\, c:C_k,\, (c,d):R_l \\ \exists C_i:\, c:C_i,\, C_i \subseteq C_k \\ \exists R_j:\, (c,d):R_j,\, R_j \subseteq R_l}} 1 \;,\quad n_{R_l.C_p} = \sum_{\substack{c,d:\, d:C_p,\, (c,d):R_l \\ \exists R_j:\, (c,d):R_j,\, R_j \subseteq R_l \\ \exists C_i:\, d:C_i,\, C_i \subseteq C_p}} 1 \;,$$

$$n_{C_k.R_l.C_p} = \sum_{\substack{c,d:\, c:C_k,\, (c,d):R_l,\, d:C_p \\ \exists C_i:\, c:C_i,\, C_i \subseteq C_k \\ \exists R_j:\, (c,d):R_j,\, R_j \subseteq R_l \\ \exists C_h:\, d:C_h,\, C_h \subseteq C_p}} 1 \;. \tag{86}$$

Accordingly, the resulting features are formulated as

$$F_{C_k.R_l} = n_{C_k.R_l} + \sum_{\substack{C_i \subseteq C_k \\ R_j \subseteq R_l}} n_{C_i.R_j} \tag{87}$$

and

$$F_{R_l.C_p} = n_{R_l.C_p} + \sum_{\substack{R_i \subseteq R_l \\ C_h \subseteq C_p}} n_{R_i.C_h} \tag{88}$$

and

$$F_{C_k.R_l.C_p} = n_{C_k.R_l.C_p} + \sum_{\substack{C_i \subseteq C_k \\ R_j \subseteq R_l \\ C_h \subseteq C_p}} n_{C_i.R_j.C_h} \;. \tag{89}$$

For attributes, related to concepts or roles, features are defined in the same way:

$$F_{C_k.A_m} = \sum_{\substack{a_i,\, a_i:A_m \,: \\ \exists c:\, (c.a_i):R_a, \\ c:C_k \lor c:C_j \subseteq C_k}} \left(a_i - \min_{A_m}\right) \cdot \left(\mathrm{rg}_{A_m}\right)^{i-1} \tag{90}$$

and

$$F_{R_l.A_m} = \sum_{\substack{a_i,\, a_i:A_m \,: \\ \exists r:\, (r,a_i):R_a, \\ r:R_l \lor r:R_j \subseteq R_l}} \left(a_i - \min_{A_m}\right) \cdot \left(\mathrm{rg}_{A_m}\right)^{i-1} \tag{91}$$

as in (78) or usage of (80) to (82) accordingly.

The mutual information based relevance rankings from section 3.1 and from chapter Chapter 4 will be calculated with

$$I(F_{C_k.R_l}; C),\; I(F_{R_l.C_p}; C) \text{ or } I(F_{C_k.R_l.C_p}; C) \;. \tag{92}$$

The features can be extended to cover even longer chains of objects and relations as well as attributes at the end of a chain.

6.1.2.2 Benefit of Combined Ontology Features over Joint Mutual Information

Mutual information itself offers a variant to consider combinations of multiple features with the joint mutual information given in (9). Therefore, the standard way to consider combinations of ontology element types would be to calculate the joint mutual information, e.g.

$$I(F_{C_k}, F_{R_l}; C), I(F_{R_l}, F_{C_p}; C) \text{ or } I(F_{C_k}, F_{R_l}, F_{C_p}; C) \tag{93}$$

instead of (92) as proposed with this thesis.

When using the joint mutual information as in (93), any numbers of specific concept occurrences of F_{C_k} (and F_{C_p}) and role occurrences of type F_{R_l} will be considered together, whether they are actually related or not. This applies, even if only concepts and roles are considered that are actually related in the ontology.

Joint mutual information of these features plainly considers the numbers of occurrences together. It does not "know" their actual relations to each other that are contained in the ontology. Using the composed features for mutual information calculation as in (92) considers only actually related occurrences of the specified concepts and roles. As a consequence, it will, by far, be a better measure of relevance of the combined ontology element types.

Additionally, ranking calculations will be faster and more accurate, because on one hand, joint mutual information takes longer to be calculated and is numerically more error-prone than single mutual information. On the other hand, rankings including the reduction of redundancy such as MRMR, MCRMR, MIFS-U, JMI and others would have to be transformed into multi feature rankings considering mutual information between feature pairs as proposed in section 0. This would tremendously increase time consumption and inaccuracy of ranking calculation.

The collision warning data set was described to contain a variety of scenes, each potentially containing several objects. These objects may appear simultaneously. Because the features of the collision warning data set can only store data of one object at a time, data of different objects is stringed together. This creates longer scenes with each object contained one after another and non-object feature data (such as ego vehicle speed, acceleration etc.) to be continued as often as there are objects in the scene. This also means that situations with different numbers of objects, e.g. vehicles or poles, cannot be distinguished, as only one object is regarded at a time.

This issue is remedied with the proposed semantic features of this chapter. In this way, even investigation of data generated by a radar sensor, which only allows poor classification, may benefit from the usage of an ontology to store detected objects and their attributes. The ontology then enables the creation of semantic features to investigate relevance or even learn classifiers as shown in Chapter 4.

6.2 Application Scenarios

The proposed method with generated ontology features enables to apply mutual information relevance calculations on ontologies. According to the relevance studies in section 4.3, a series of measurement data gathered in an ontology is required in advance. This ontology (ontological time series) contains classified objects, relations among them and attributes. Instead of using plain measurement data, ontology elements may also be labeled by experts upon raw data, such as video sequences, to provide classifications and relations as correct as possible.

Necessary for ontology feature generation, as proposed in section 6.1, is an underlying taxonomy of concepts and a hierarchy of roles as well as classified and related measurement data.

The feature rankings will then provide information about relevance of concepts and roles and what roles between concepts already assigned in the data are useful for a given goal function.

For example, it is useful to know the current type of the road, but perhaps it may not be gained directly from sensor information. Ontology feature rankings may then be used to find out ontology element types relevant to determine the type of road currently driven on, e.g. motorway, rural or urban. This even may include attributes such as velocities, lane widths, road sign sizes and so on. Necessary, however, is a sufficient amount of labeled data containing the actual road type and labeled semantic information.

In general, ontology feature rankings may provide information on the following items:

- The classification tasks a sensor data fusion should concentrate on.
- The classifications that should be especially accurate and thus should be improved concerning object fusion and detection.
- The roles, which are of importance and the according relations, which should be set by sensor data fusion.
- The ontology elements that may be omitted to reduce the size of the ontology and speed up reasoning
- The most beneficial focus research and development effort on sensor data fusion and ontology engineering should lie on.
- An effective distribution of resources within sensor data fusion calculations.
- The quality of measurement data and object classification.
- A benchmark of implemented sensor data fusion methods with respect to reference sensor data.

The aspects about mutual information feature selection as described in section 3.1.3 as well as chapter Chapter 4, especially in 4.1, 4.3 and 0 apply accordingly.

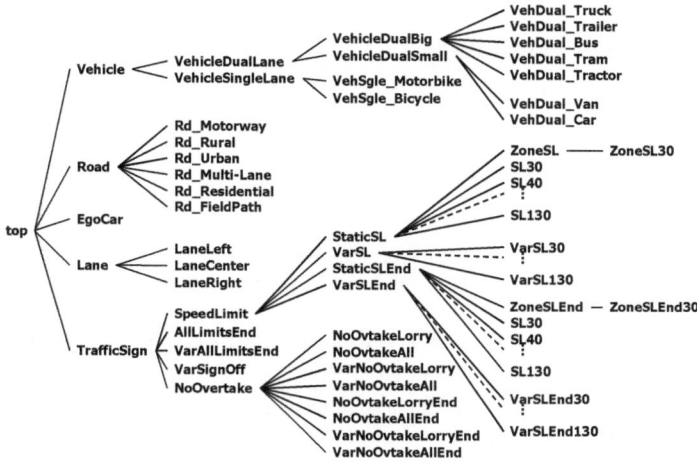

Fig. 54: Taxonomy of concepts used to label objects in real world video recording and measurement sequences underlying the exemplary application of an ontology feature ranking.

6.3 Sample Application of an Ontology Feature Ranking

6.3.1 Data and Labeling Description

To show the application of an ontological feature ranking in principle, it was carried out on manually labeled video sequences. For object labeling, the taxonomy shown in **Fig. 54** was used. Sequences had already been partially labeled for object detection purposes, considering traffic participants, lanes, the ego vehicle relative to lanes and the traffic participants relative to the ego vehicle. Unfortunately, traffic participants were not labeled relative to lanes. Because labeling traffic participants is highly time-consuming and requires a lot of labeling personnel, relabeling could not be carried out for this thesis. However, in addition, traffic signs were labeled throughout the exemplary picked sequences. The labeled types of traffic signs are subsumed by TrafficSign in the taxonomy.

The goal of this example application is to show relevance of ontological situation features with respect to the labeled road type. The ranking should provide information to what ontological, semantic features provide high information to decide about the current type of road, e.g. motorway, rural road or urban road or multi-lane road (see Road in **Fig. 54**). The multi-lane road type helps to distinguish between motorways and bigger rural or urban roads, which are not motorways. Scenes were selected in a manner to represent the common 1/3-mix (see **Fig. 55**).

Overall, labeled data comprises 76 scenes with 56,048 data points (labeled frames), representing about 2,242 *s*. In average, this means an approximate scene length of 29.5 *s*. Unfortunate-

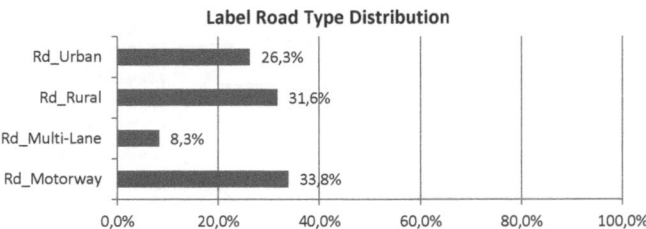

Fig. 55: Distribution of numbers of data points of the different road types from the video labeling data set underlying the mutual information based ranking calculation of ontology features.

ly, the scenes are relatively short. This is, because scenes were originally created for plain object detection, not to create consecutive time series.

As a consequence for this example application of ontological feature ranking, traffic signs could not be labeled in long scenes as desired. The distance traveled in one scene, e.g. at a speed of 50-60 *km/h* for 30 *s* is roughly 500 *m*. That is a problem, because traffic signs do not occur as often as, e.g. traffic participants or lanes, which are always present. Hence, the currently valid traffic sign is mostly unknown at scene beginning and can only be labeled once a traffic sign occurs. This may happen very late, leaving only a small fraction of the scene to be labeled. A lot of traffic sign information is lost or unknown, respectively, for that reason. In the rankings, traffic sign relevance should end up much higher, when labeled consecutively, than can result in this example.

6.3.2 Ontology Feature Ranking Results on Labeled Data

Table 8 shows an excerpt of the calculated Maximum Relevance ranking for the created ontological features. The full ranking is provided in **Table 10** in 0. Feature names starting with a capital letter are concepts, features starting lowercase are roles and those divided by points are features about chains of concepts and roles. Overall, 289 features were created evaluating the time-series ontology[59].

It is obvious that the largest part of the relevance to decide about the type of road is provided by the number of lanes in a scene. The relative mutual information of the Lane feature w. r. t. the road type class is 34.68%.

It is followed directly by the feature indicating existence of a left lane, with a relative mutual information of the LaneLeft feature of 31.07%. That is, because a left lane is virtually always existent, except, if there is only one lane. In the time-series ontology, which is derived from labeled scenes, the road is always rural or urban, if there is no left lane and, in slightly more cases, it is urban rather than rural. Hence, it is already a very good indicator to neglect motorway and multilane and even put a tendency towards an urban road classification.

---

[59] For better readability, some features are left out that are completely redundant with illustrated features.

Both features each make up around one third of the MI with the class. Keep in mind, however, that information of both features might overlap.

The isPart-relation is the third most relevant feature. This is, because both lanes as well as traffic signs can have the isPart-relation to Road. Note, that in the labeled data there is always only one road and hence it has the information and mutual information zero. It is, because traffic signs are of less relevance in the labeled data, that the isPart-relation is less relevant than the concept feature of Lane.

A very interesting benefit of the ontological feature selection can be seen and discussed at the example of the EgoCar.isOn.LaneRight-feature. Ranked at position 10, this feature has comparably high relevance with 16.39% relative MI. Looking at its components EgoCar and LaneRight, their information (entropy) and MI are exactly zero. That is, because both EgoCar and LaneRight are always existent. However, the conjoined feature reaches a high relevance, whereas the feature for the isOn-relation only follows at position 77 with roughly 3.93% relative MI. This serves as a good illustration of the benefit to use features considering semantics in the ontology. Calculation of the joint mutual information of all 3 single features with the class would erroneously consider any isOn-occurences together with EgoCar and LaneRight, not only the semantically connected occurrences and would most probably come along with some numerical problems calculating the JMI. Moreover, the JMI of only EgoCar and LaneRight would be zero as well.

In addition to the relative position of the ego vehicle on the road (kind of lane used), it is relatively important to know the number of vehicles driving on a lane right of the ego vehicle (rank 15 with 12.19%). In comparison, the number of vehicles left of the ego vehicle ranks down on position 72 with only 4.03%.

Note, that the number of vehicles on a lane relatively positioned to the ego lane (feature Vehicle.isRelPos.EgoCar) ranks slightly higher than the total number of vehicles (feature Vehicle), which includes any seen vehicles that may also be located off the road (e.g. vehicles on a parking place). Overall, the total number of vehicles has only medium relevance with a relative MI of 7.55%.

Besides traffic objects, traffic signs were labeled in all scenes. Unfortunately, as explained above, traffic signs could not be labeled consecutively and hence only small parts of the scenes and not even all scenes do have labeled traffic sign data. As a consequence traffic signs must rank very low (all specific types less than 4.62% relative MI), because their entropy must be low preliminarily. Only the overall number of traffic signs in the feature TrafficSign, considering any sign, does have noticeably higher relative MI of 6.88%.

The effective mutual information of traffic signs and specific types, assuming fully labeled scenes, is expected to be much higher. For example, SL120, SL130 and VarSLxx usually (perhaps even only) occur on motorways, whereas SL80 is typical for construction sites on motorways, SL70 and SL90 are typical on rural roads and SL50 and SL30 in cities (urban roads).

The result in this case, however, are traffic signs ranked at the following positions (in parentheses): SL70 (62), SL100 (110), SL90 (126), SL80 (134), SL30 (141), SL50 (153), VarSL100 (170), VarSL120 (194) and so on. Traffic signs with end of speed limits rank even lower.

Table 8:   Extract of the maximum relevance ranking of generated features from the video labeling sequence ontologies (time series ontology). The full ranking is provided in **Table 10** in 0.

Pos	Semantic Feature F	Entropy H(F)	1.8584 I(F;C)	Label Road Type (Rd_...) I(F;C)/H(C)	I(F;C)/H(F)
1	Lane	1.5794	0.6445	34.68%	40.80%
4	LaneLeft	0.9996	0.5773	31.07%	57.76%
7	isPart	2.2481	0.3563	19.17%	15.85%
10	EgoCar . isOn . LaneRight	0.9101	0.3046	16.39%	33.47%
15	Vehicle . isRightOf	1.6155	0.2265	12.19%	14.02%
17	Vehicle . isRightOf . EgoCar	1.6155	0.2265	12.19%	14.02%
33	Vehicle . isRelPos . EgoCar	3.0066	0.1507	8.11%	5.01%
37	Vehicle	3.2421	0.1403	7.55%	4.33%
40	TrafficSign	1.6237	0.1278	6.88%	7.87%
58	SpeedLimit	1.3967	0.0859	4.62%	6.15%
62	SL70	0.5968	0.0846	4.55%	14.17%
72	Vehicle . isLeftOf . EgoCar	1.9624	0.0749	4.03%	3.82%
77	isOn	0.4875	0.0731	3.93%	14.99%
110	SL100	0.4324	0.0562	3.03%	13.01%
123	VarSL	0.1936	0.0480	2.58%	24.78%
126	SL90	0.0941	0.0445	2.40%	47.33%
134	SL80	0.1833	0.0376	2.03%	20.54%
141	SL30	0.1340	0.0320	1.72%	23.89%
153	SL50	0.1951	0.0289	1.55%	14.79%
170	VarSL100	0.1221	0.0264	1.42%	21.63%
194	VarSL120	0.1015	0.0209	1.13%	20.63%
283	EgoCar	0.0000	0.0000	0.00%	NaN
284	Road	0.0000	0.0000	0.00%	NaN
285	LaneRight	0.0000	0.0000	0.00%	NaN
288	EgoCar . isOn . Road	0.0000	0.0000	0.00%	NaN

6.4 Summary

This chapter combined both approaches of the two previous chapters mutual information based situation feature relevance and knowledge-based situation description. As a consequence, it is now possible to determine the relevance of element types of an ontology, which can be interpreted as situation features, with respect to some target application or target measure.

Ontologies were found suitable for a holistic description of traffic situations, but may yet be oversized or undersized. The method presented and introduced in this chapter is hence useful to determine, what information within the ontology could be omitted or if (and to some extent how much) the ontology does not provide sufficient information for the target application.

This is done for elements, formulated and formalized as semantic features, such as concepts and roles and attribute types, but also combined ontology element types, i. e. ontology subgraphs. These semantic features are derived from ontological time series. They are then used for mutual information calculation with respect to a target function (or class), which also may be a semantic feature from the ontology, and mutual information based feature selection can be carried out just as presented in the previous chapters.

Knowing the relevance of ontological, semantic features in the consequence facilitates effective and efficient ontology engineering and to creating lean, but capable and expressive knowledge-based systems, especially driver assistance systems. What information can be derived with the aim to achieve this in respect to semantic feature quality and selection or research and development focus was described in this chapter, particularly in the domain of advanced driver assistance systems.

Performing mutual information calculations on semantic features requires a sufficient amount of labeled data, preferably taken from real world measurements, which were semantically labeled.

Application and expressiveness was shown in principle in a way that the method was carried out on manually labeled video sequences from real world driver assistance test runs. These labels considered traffic participants, lanes, the ego vehicle, relations to lanes and other traffic participants as well as traffic signs. The goal of this example application was the relevance of such kind of ontological, semantic situation features with respect to a labeled road type.

Chapter 7

Conclusion

With the European Commission among other international agencies aiming to significantly reduce the number of traffic fatalities ongoing and in the future as well as to improve the performance of traffic systems and participants intensive research on Driver Assistance Systems is part of today's science. These systems support the driver in critical situations or intervene in the driving process to avoid accidents or to reduce their severity. In addition, comfort assistance functions are increasingly making driving more convenient.

In this sense, advanced driver assistance systems (ADAS) help the driver with information or by performing supportive or autonomous actions. This demands an extensive comprehension of the vehicle environment and complex situations, potentially even unknown situations. Therefore, for further situation interpretation, a generic method for situation description is needed.

The comprehension of a traffic situation transforms perceived raw information into interpretable information. This is the basis for future projection, decision making and action performing, such as navigating, maneuvering and driving control.

7.1 Contributions

The result of this thesis is a framework paired with a profound theoretical discussion of a generic traffic situation description capable of supplying various ADAS with relevant information about the current driving and traffic situation of the ego vehicle and its environment. With this information ADAS will be enabled to perform reasonable functions and actions and approach visionary goals such as injury and accident free driving, substantial assistance in arbitrary situations up to even autonomous driving.

It is made feasible to assess more complex situations compared to state of the art assistance systems, potentially including even unknown situations demanding for extensive comprehension of the vehicle environment and the current situation. The feasibility is empirically proven on exemplary assistance functions in a realtime 3D driving dynamics simulation environment.

Situation Feature Selection

As part of describing a situation, it is of particular interest, what information is actually relevant to target applications. For ADAS, a number of partially redundant sensors is used combined with further feature generation algorithms to generate a variety of measurement signals. *Mutual Information* (MI) is a well-established statistical tool for feature selection and feature comparison, which does not preliminarily need a classifier or other machine learning algorithms for relevance calculation.

Knowledge-based Traffic Situation Description

Most complex traffic situations seem to be those at intersections. Their understanding is influenced by a variety of object and relation types such as intersecting roads with lanes and markings, allowed and forbidden paths, vehicles coming from different directions and different kinds of road signs. Their constellation directly influences traffic rules which apply and, accordingly, the assessment of allowed actions, expected behavior and impact of traffic participants among each other.

Ontologies are a foundation for knowledge representation and provide a formalism to structure objects, their relations and attributes and for performing logic reasoning with them. Therefore, ontologies are well suited for modeling these multi-object traffic situations and for performing logic reasoning to check consistency of its knowledge and to reason about object types, relations and to e.g. apply traffic rules.

Description logic (DL) is a language for building ontologies and in most cases, depending on the dialect applied, it allows for decidable, complete and terminating algorithms.

Key Research Contributions within this Thesis

Three key contributions are provided by this thesis in the context of advanced driver assistance systems, more precisely to efficiently describe the current traffic situation a vehicle is part of:

1. The empirical proof that *mutual information based feature selection* serves to effectively evaluate the relevance of or select *traffic situation features*, especially those provided by ADAS sensors and ADAS sensor data fusion.

2. Theoretical discussion and empirical proof that *description logic based ontologies* are suitable to describe *complex traffic situations* such as those at intersections, including the evaluation of traffic rule compliance.

3. Theoretical discussion and empirical proof that *mutual information based feature selection* may be utilized on *semantic features* contained in *description logic based ontologies* for a more comprehensive and lean traffic situation description.

Contribution 1: Mutual Information Based Traffic Situation Feature Selection

Mutual information already serves as a widely used tool for comparison of features and feature selection, especially to support machine learning algorithms in the field of artificial intelligence. This thesis presents various applications in the development of driver assistance systems, especially to manage situational complexity and the variety of vehicle and environment measurement signals.

It demonstrates the successful implementation using large amounts of measurement data. The applied methods are particularly suitable due to their ease of use and automated handling of many features, especially for comparison and for determining the importance of features for specific driver assistance system functions. It could be demonstrated that, for the presented application for collision assistance systems, superiority was achieved over the feature selection by experts working in serial function development (see also publication [Hülsen et al., 2010]).

In addition, a methodical improvement to existing methods including formalization and error calculations is given and empirically shown with this thesis.

As an important contribution, an approach and its interpretation for the usage of mutual information to determine the relevance of the history of situation features is developed. This approach considers past feature values laying back a certain amount of time.

It provides mutual information calculations based on shifted time series values of features with respect to a target function. The method may be used to derive new features, identify important time delays or to provide predictive signals. The approach and applicatioon is illustrated by simulation results on measurement data from vehicle endurance runs.

In the near future, studies will evaluate to what extent the presented methods may support system development of serial products and how they actually contribute to functional improvement.

Contribution 2: Description Logic Based Traffic Situation Description for ADAS

This thesis is widely concerned with the investigation and validation of applicability and capability of ontology based traffic situation description. An ontology able to model traffic situations at complex intersections and to enable reasoning about traffic rules for involved vehicles is introduced. In contrast to reference work this ontology is not aimed to provide detailed geometric modeling with extensive but slow reasoning but to be lean to facilitate fast reasoning close to real-time with a framework to allow modular ontology building for different DAS agents with different reasoning goals. This ontology not only accounts for plain knowledge representation but for consistency checking of e.g. sensor information as well as knowledge interpretation and augmentation by applying reasoning based on description logic.

Knowledge engineering for complex traffic situations such as those at complex intersections with an arbitrary number of roads, lanes, driving directions, allowed driving paths, traffic signs and / or lights was performed and is proven to be feasible. Furthermore, the developed description logic based ontology allows logic reasoning of traffic rules.

Logic reasoning of traffic rules may be performed in different modes according to the state or assumption of certainty or uncertainty. Given, driving paths of all or some vehicles are known, traffic rules will be reasoned with exact results for these vehicles. Given driving paths are completely unknown, all possibilities will be considered and contained in the reasoning result. Provided, allowed driving paths are known, e.g. by road markings or traffic signs, only the allowed paths will be considered for traffic rule reasoning if opted for. Should traffic signs at an intersection be partially unrecognized, undetected or occluded, proper types of traffic signs will be reasoned by the ontology if logically possible.

Both the developed ontology for complex traffic situations and the inclusion of logic reasoning on traffic rules have been published in [Hülsen et al., 2011b].

Additionally, this thesis provides a proof-by-implementation for logic-based situation description for real-time execution of driver assistance functions. An asynchronous real-time framework is used, especially designed for the herewith proposed ontological situation description. It is usable for arbitrary DAS functions; shown on exemplary DAS functions (see also publication [Hülsen et al., 2011a] and associated master theses [Hesser, 2011, Spinda, 2011]).

An open issue with ongoing research is coping with uncertainty in combination with knowledge-based approaches. Early approaches are too slow for implementation, especially with the number of objects and relations required for the proposed, generic situation description. This thesis introduces and discusses a concept of handling and incorporating different types of sensory uncertainty with the deterministic ontological situation description provided in this thesis to facilitate some sensory uncertainty handling.

Contribution 3: Mutual Information Based Selection of Ontology Features

Finally, this thesis defines semantic features derived from ontological content to perform mutual information based feature selection on ontology elements and to determine their relevance with respect to some given target application. Execution of this approach on some experimental data shows its applicability in principle.

These introduced semantic features have a significant benefit over traditional features. When creating these features their semantic relation to other ontological entities as well as hierarchical relationships are taken into account. In this way, relevance of attributes of a specific type, which belong to a certain ontological class or relation, may be retrieved as well as the overall relevance of the attribute. Moreover, relevance information of complete sub-graphs of particular interest within an ontology may be provided. This has comprehensive added-value to using multivariate mutual information instead, because occurrences and structure of such sub-graphs are taken into account, not only plain occurrence of its components.

7.2 Outlook

Research on advanced driver assistance systems is a large area of functionality, functional improvements and performance yet discoverable. To date, sophisticated systems are already available on the market or close to market penetration. However, on its way to fully autonomous driving, intra-communication infrastructure and inter-communicating traffic participants as well as comprehensive driver support in any situation, huge potential for research persists.

This thesis provides a profound and empirically proven basis and discussion for a generic situation description. This section will briefly point out, what further research and development underpinned by this thesis is supposed to be beneficial.

Extensive use of mutual information

Today, vast amounts of data are made available through vehicle endurance and test runs. Currently these runs and their recorded data are aimed for system testing, approval and release as well as the verification of a certain Automotive Safety Integrity Level (ASIL), for which a certain statistical failure rate has to be empirically and/or theoretically proven.

The amounts of data are so huge, that even multi-dimensional statistical calculations can be carried out (for example data covers several hundred thousands of kilometers with each data point being recorded at each tenth of a second or less). During research work on mutual information applications in this thesis, several hundred kilometers were already available with consistent data able to be processed. It was shown in this thesis that even on this amount of data, mutual information applications can be superior compared to expert knowledge.

With the amounts of data available in the near future, extensive application of statistical calculation and learning methods are enabled, i.e. automated feature selection, feature evaluation, feature generation and machine learning. These methods do have tremendous potential to support or to a certain extent replace current system development by rather using real-world field statistics than the partially limited view of plain expert knowledge.

Fully probabilistic logic based generic situation description and interpretation

This thesis particularly deals with providing a logic-based generic situation description, delineated from sensory perception and from situation interpretation and decision making. It is discussed that the pure logic-based generic situation description does not incorporate any probabilistic formulations or probabilistic reasoning. However, probabilistics and uncertainty in general are inherent to perception, interpretation as part of situation awareness and decision making.

A research area recently emerging and concretizing is probabilistic logic reasoning. First implementations are proven to be successful and comprehensiveness, capabilities and computational efficiency are steadily increasing. Probabilistic logic reasoning and, hence, ontologies combined with probabilistic formulations have the potential of providing a formal foundation to model and reason about entire situation awareness including all of sensory perception, situation description, situation interpretation and projection, behavior modeling and decision making or combinations of parts of these.

Besides the logic-based generic situation description, this thesis discusses a basic but efficient approach to make some sensory uncertainty handling feasible as long as more sophisticated and comprehensive methods are not. Probabilistic logic reasoning methods do have the potential to directly include sensory uncertainty in the logic-based generic situation description. No further particular treatment would be necessary with proper expressiveness of the probabilistic logic reasoning methods, once available. Sentences or answers of the generic situation description would always directly be provided together with a measure of uncertainty or belief in its truth, respectively.

The same basically applies for among others situation interpretation, behavior modeling and decision making. However, finding an approach available for a wide range of applications may turn out to be a challenge, as these components of driver assistance systems are in general very application specific. However, especially combined with comprehensive machine learning techniques, potentially even considering semantic features as proposed with this thesis, exploration may be of great research interest.

To date, necessary items to enable fruitful research and results within the domain of ADAS are:

- Sufficient probabilistic and ontological logic language / expressiveness of the underlying probabilistic logic description logic. Particularly the description logic dialect used in this thesis has to be fully supported.

- An comprehensive and understandable probabilistic and logic modeling and reasoning of the ontology. A large advantage of the *generic* logic based situation description described in this thesis is its transparency concerning its modeled background knowledge. Axioms and rules are formulated to be understandable by a human expert. The complexity of a generic situation description may become unwieldy with too complex probabilistic formulations in addition. However, this problem could be solved with powerful machine learning algorithms, which make use of large available amounts of data for model learning and model verification.

- A feasible dimension of time consumption of considered reasoning algorithms has to be provided. This concerns both the order of reasoning algorithms (in general for this domain of algorithms it is exponential) as well as the efficiency of modeling and overal time consumption. As target, the generic probabilistic logic based situation description has to be processed in real-time. The order of dimension is of particular interest to assure scalability of the situation description for different complexities of situations.

Abduction and learning

Besides the future necessity of probabilistic formulations or handling of uncertainty within the logic-based situation description, respectively, abduction reasoning and logic learning is another field of extensive and promising research. First approaches for abduction reasoning exist (see e.g. [Llinas et al., 2004]).

Abduction is part of logic reasoning and deals with creating new knowledge / new instances from abductive rules, perhaps even creating new classes or roles from abductive learning. Usually, because there may be many possibilities for abductive creation at once, these rules are paired a measure for degrees of belief of their truth. Hence, there is a link to probabilistic logic reasoning.

Examples in the domain of traffic situations for abductive rules would be:

- E.g. creating a road if traffic signs are detected (road occluded)

- E.g. assuming vehicles at occluded intersections

- E.g. creating remaining traffic signs if certain other ones are detected[60]

Current, early abduction approaches are in an early state of research and very time consuming. However, particularly together with probabilistic logic reasoning, potential for the domain of advanced driver assistance systems and a generic situation awareness is strong.

---

[60] Note, that for reasoning about unknown traffic signs included in the generic situation description proposed with this thesis, all traffic signs have to be instantiated prior to ontological reasoning by external algorithms. However, their classification may be left unknown and will be reasoned within the ontology.

Early implementation

This thesis introduces a framework for a generic situation description including approaches to determine and investigate feature relevance. These methods have been tested on real-world test cases run in simulation environments. It was desirable to perform testing with in-vehicle-application. However, even research vehicle sensory equipment, system integration and computing equipment was not available to the necessary extent during research, development and testing phase of this thesis. Singular equipment does exist, which was used for plausibility of specific parts of the scope of this thesis. High level information, among many others such as traffic light and sign recognition, lane and road marking detection, road, lane and intersection detection, detailed object detection including proper classification, was not yet available and therefore has been simulated to prove the capabilities of the methods and models introduced with this thesis.

However, this thesis provides a foundation for future advanced driver assistance development. Currently, sensory equipment and computing power within vehicles is rapidly improving and connecting, so that in the near future, conditions for in-vehicle application of this thesis' results are assumed to be given.

Appendix A Full Ranking Tables

A.1 Maximum Relevance Ranking of two Joint Features

Table 9: Extract of the maximum relevance ranking with 2 joint features for de-escalating maneuvers (missing ranking positions are due to nondisclosure).

Pos	Features / Signals *F,G*	Entropy $H(F)+H(G|F)$	0.3713 $I(F,G;C)$	$H(C)$ (De-escalation class) $I(F,G;C)/H(C)$	$I(F,G;C)/H(F,G)$
1	Object_DyCirc , TTC	1,4707	0,3272	88,12%	22,25%
2	Object_Dy , TTC	1,8088	0,3233	87,05%	17,87%
3	Object_Dx , TTC	2,5172	0,3216	86,60%	12,78%
5	Object_VxRel , TTC	1,7918	0,3165	85,24%	17,67%
8	Object_Dx , Object_VxRel	2,7024	0,3121	84,06%	11,55%
10	EgoVx , TTC	6,2297	0,3112	83,80%	4,99%
12	Object_Vx , TTC	1,7846	0,3099	83,45%	17,36%
13	Object_VyRel , TTC	1,3637	0,3095	83,35%	22,70%
15	GasPedal , TTC	5,0623	0,3056	82,29%	6,04%
18	TTC , WheelAngle	2,5999	0,3035	81,74%	11,68%
19	EgoAx , TTC	4,0768	0,3034	81,69%	7,44%
22	Object_AxRel , TTC	1,7077	0,3012	81,13%	17,64%
23	EgoAy , TTC	2,6021	0,3010	81,05%	11,57%
25	GasPedalDt , TTC	3,4769	0,3003	80,87%	8,64%
27	TTC , WheelAngleDt	1,9526	0,2992	80,58%	15,32%
28	Object_Vy , TTC	1,1082	0,2986	80,40%	26,94%
37	DirIndL , TTC	1,0601	0,2893	77,91%	27,29%
38	DirIndR , TTC	1,0589	0,2890	77,84%	27,29%
39	TTC	0,8120	0,2880	77,55%	35,46%
42	Object_DyCirc , Object_VxRel	1,8137	0,2582	69,52%	14,23%
44	Object_Dy , Object_VxRel	2,1046	0,2574	69,33%	12,23%
46	GasPedal , Object_VxRel	5,4991	0,2550	68,66%	4,64%
48	Object_Vx , Object_VxRel	1,9809	0,2495	67,19%	12,60%
49	EgoVx , Object_VxRel	6,4228	0,2495	67,19%	3,88%
50	Object_VxRel , WheelAngle	3,0649	0,2490	67,05%	8,12%
51	Object_VxRel , Object_VyRel	1,7227	0,2485	66,91%	14,42%
53	EgoAx , Object_VxRel	4,5306	0,2467	66,44%	5,45%
54	Object_AxRel , Object_VxRel	1,9886	0,2461	66,28%	12,38%
58	EgoAy , Object_VxRel	3,0688	0,2437	65,64%	7,94%
61	Object_Dx , Object_Vx	2,5750	0,2412	64,95%	9,37%
64	GasPedalDt , Object_VxRel	3,9446	0,2383	64,18%	6,04%
65	Object_VxRel , WheelAngleDt	2,4248	0,2366	63,71%	9,76%
68	Object_VxRel , Object_Vy	1,5220	0,2325	62,61%	15,27%
74	EgoVx , Object_Dx	7,1517	0,2259	60,83%	3,16%
75	DirIndR , Object_VxRel	1,5270	0,2221	59,81%	14,54%
76	DirIndL , Object_VxRel	1,5306	0,2213	59,60%	14,46%
79	Object_Dx , Object_DyCirc	2,4049	0,2205	59,39%	9,17%
80	Object_VxRel	1,2871	0,2187	58,89%	16,99%
81	Object_Dx , Object_Dy	2,5577	0,2152	57,96%	8,41%
87	Object_Dx , Object_VyRel	2,3769	0,2074	55,86%	8,73%
97	Object_AxRel , Object_DyCirc	1,6698	0,1970	53,07%	11,80%
105	Object_Dy , Object_Vx	1,9823	0,1912	51,49%	9,65%
107	Object_DyCirc , Object_Vx	1,7144	0,1907	51,37%	11,13%
109	Object_AxRel , Object_Vx	1,8711	0,1907	51,35%	10,19%
110	GasPedal , Object_Dx	6,1168	0,1902	51,21%	3,11%

112	Object_Dy , Object_VyRel	1,6310	0,1891	50,94%	11,60%
116	Object_AxRel , Object_Dy	1,9464	0,1855	49,96%	9,53%
119	Object_DyCirc , Object_VyRel	1,2804	0,1848	49,77%	14,43%
126	EgoVx , Object_Dy	6,5305	0,1822	49,06%	2,79%
128	EgoAy , Object_Dx	3,7144	0,1815	48,87%	4,89%
134	EgoVx , Object_DyCirc	6,2242	0,1807	48,66%	2,90%
136	EgoAx , Object_Dx	5,1722	0,1805	48,61%	3,49%
139	Object_AxRel , Object_Dx	2,5618	0,1798	48,42%	7,02%
140	Object_Vx , Object_VyRel	1,6248	0,1797	48,40%	11,06%
148	GasPedalDt , Object_Dx	4,5698	0,1776	47,84%	3,89%
149	Object_Dy , Object_DyCirc	1,5666	0,1775	47,79%	11,33%
153	Object_Dx , WheelAngle	3,7020	0,1768	47,61%	4,78%
155	Object_AxRel , Object_VyRel	1,5771	0,1767	47,60%	11,21%
170	EgoAx , Object_DyCirc	4,0948	0,1724	46,43%	4,21%
185	Object_Dx , Object_Vy	2,1588	0,1706	45,95%	7,90%
191	Object_Dx , WheelAngleDt	3,0813	0,1687	45,44%	5,48%
195	EgoAy , Object_Dy	2,9340	0,1683	45,33%	5,74%
196	Object_Dy , WheelAngle	2,9414	0,1673	45,05%	5,69%
198	Object_DyCirc , WheelAngleDt	1,9640	0,1667	44,89%	8,49%
200	EgoAx , Object_Dy	4,4345	0,1665	44,84%	3,76%
204	GasPedal , Object_DyCirc	5,0687	0,1660	44,72%	3,28%
205	Object_DyCirc , WheelAngle	2,5956	0,1660	44,70%	6,39%
207	EgoAy , Object_DyCirc	2,6033	0,1659	44,69%	6,37%
220	EgoVx , Object_VyRel	6,1120	0,1615	43,48%	2,64%
221	GasPedal , Object_Dy	5,3983	0,1612	43,41%	2,99%
224	GasPedalDt , Object_DyCirc	3,4955	0,1600	43,08%	4,58%
233	Object_Dy , WheelAngleDt	2,3122	0,1581	42,58%	6,84%
235	Object_DyCirc , Object_Vy	1,0538	0,1579	42,51%	14,98%
249	GasPedalDt , Object_Dy	3,8348	0,1550	41,73%	4,04%
250	Object_Dy , Object_Vy	1,3626	0,1547	41,66%	11,35%
265	EgoVx , Object_AxRel	6,4436	0,1514	40,78%	2,35%
267	DirIndR , Object_DyCirc	1,0662	0,1511	40,69%	14,17%
270	DirIndR , Object_Dx	2,1913	0,1504	40,50%	6,86%
271	DirIndL , Object_DyCirc	1,0682	0,1502	40,45%	14,06%
278	EgoAx , Object_VyRel	3,9797	0,1492	40,18%	3,75%
279	DirIndL , Object_Dx	2,1915	0,1488	40,07%	6,79%
280	Object_DyCirc	0,8201	0,1487	40,04%	18,13%
288	Object_Dx	1,9451	0,1464	39,43%	7,53%
290	GasPedal , Object_VyRel	4,9582	0,1459	39,30%	2,94%
291	EgoAy , Object_VyRel	2,4541	0,1459	39,29%	5,94%
300	DirIndR , Object_Dy	1,4113	0,1432	38,57%	10,15%
301	DirIndL , Object_Dy	1,4126	0,1420	38,23%	10,05%
303	Object_VyRel , WheelAngle	2,4709	0,1417	38,16%	5,73%
306	Object_Dy	1,1653	0,1401	37,73%	12,02%
308	Object_VyRel , WheelAngleDt	1,8446	0,1389	37,41%	7,53%
309	GasPedalDt , Object_VyRel	3,3811	0,1389	37,40%	4,11%
316	EgoAx , Object_AxRel	4,1477	0,1361	36,65%	3,28%
319	Object_Vy , Object_VyRel	0,9003	0,1345	36,23%	14,94%
325	Object_AxRel , Object_Vy	1,3087	0,1325	35,69%	10,13%
329	EgoAy , Object_AxRel	2,8396	0,1291	34,76%	4,55%
330	Object_AxRel , WheelAngle	2,8349	0,1290	34,74%	4,55%
334	GasPedal , Object_AxRel	5,2661	0,1267	34,12%	2,41%
335	DirIndR , Object_VyRel	0,9508	0,1266	34,10%	13,32%
336	DirIndL , Object_VyRel	0,9526	0,1261	33,97%	13,24%
341	Object_VyRel	0,7040	0,1245	33,53%	17,69%
345	EgoVx , Object_Vx	6,2601	0,1234	33,23%	1,97%
347	Object_AxRel , WheelAngleDt	2,1913	0,1224	32,95%	5,58%

349	GasPedalDt , Object_AxRel	3,7128	0,1210	32,60%	3,26%
376	DirIndR , Object_AxRel	1,2929	0,1088	29,29%	8,41%
377	DirIndL , Object_AxRel	1,2949	0,1085	29,21%	8,38%
383	EgoAx , Object_Vx	4,3195	0,1078	29,04%	2,50%
390	Object_AxRel	1,0471	0,1064	28,67%	10,17%
398	Object_Vx , WheelAngle	2,8433	0,1028	27,68%	3,61%
403	EgoAy , Object_Vx	2,8342	0,1020	27,48%	3,60%
405	GasPedal , Object_Vx	5,2687	0,1014	27,31%	1,92%
415	Object_Vx , Object_Vy	1,2566	0,0982	26,45%	7,82%
417	Object_Vx , WheelAngleDt	2,2100	0,0968	26,08%	4,38%
435	GasPedalDt , Object_Vx	3,7223	0,0915	24,64%	2,46%
456	DirIndR , Object_Vx	1,3106	0,0828	22,29%	6,32%
459	DirIndL , Object_Vx	1,3115	0,0821	22,10%	6,26%
460	EgoVx , GasPedal	9,4031	0,0818	22,02%	0,87%
461	EgoVx , Object_Vy	5,7674	0,0817	22,00%	1,42%
465	Object_Vx	1,0665	0,0803	21,62%	7,53%
487	EgoAy , Object_Vy	2,1237	0,0707	19,03%	3,33%
491	EgoAx , EgoVx	8,4727	0,0690	18,59%	0,81%
496	EgoAx , Object_Vy	3,6210	0,0673	18,13%	1,86%
502	Object_Vy , WheelAngle	2,1342	0,0666	17,94%	3,12%
523	EgoAy , EgoVx	7,0989	0,0617	16,62%	0,87%
527	GasPedal , Object_Vy	4,5984	0,0613	16,51%	1,33%
533	Object_Vy , WheelAngleDt	1,4913	0,0597	16,09%	4,01%
535	EgoVx , WheelAngle	7,0202	0,0596	16,06%	0,85%
537	EgoVx , GasPedalDt	8,0073	0,0594	15,99%	0,74%
550	GasPedalDt , Object_Vy	3,0245	0,0535	14,42%	1,77%
558	EgoVx , WheelAngleDt	6,4717	0,0495	13,33%	0,76%
568	DirIndR , Object_Vy	0,5879	0,0462	12,44%	7,86%
570	DirIndL , Object_Vy	0,5888	0,0458	12,33%	7,78%
575	Object_Vy	0,3399	0,0443	11,93%	13,03%
577	EgoAx , WheelAngle	5,0335	0,0439	11,82%	0,87%
587	GasPedal , GasPedalDt	6,6023	0,0388	10,45%	0,59%
588	EgoAx , EgoAy	5,0428	0,0383	10,32%	0,76%
590	EgoAx , WheelAngleDt	4,3867	0,0373	10,05%	0,85%
592	EgoAx , GasPedal	7,0308	0,0370	9,97%	0,53%
594	DirIndL , EgoVx	5,6924	0,0365	9,83%	0,64%
595	DirIndR , EgoVx	5,6906	0,0355	9,57%	0,62%
602	EgoAy , GasPedal	6,0329	0,0341	9,18%	0,57%
606	GasPedal , WheelAngle	6,0102	0,0332	8,94%	0,55%
609	EgoAx , GasPedalDt	5,8465	0,0328	8,83%	0,56%
610	EgoVx	5,4548	0,0326	8,78%	0,60%
611	WheelAngle , WheelAngleDt	2,7994	0,0325	8,74%	1,16%
619	EgoAy , WheelAngle	3,3218	0,0301	8,11%	0,91%
623	GasPedal , WheelAngleDt	5,4069	0,0297	8,01%	0,55%
626	EgoAy , WheelAngleDt	2,8888	0,0294	7,93%	1,02%
632	GasPedalDt , WheelAngle	4,4515	0,0268	7,23%	0,60%
637	EgoAy , GasPedalDt	4,4632	0,0256	6,89%	0,57%
644	GasPedalDt , WheelAngleDt	3,8231	0,0238	6,42%	0,62%
652	DirIndR , EgoAx	3,5264	0,0220	5,93%	0,62%
653	DirIndL , EgoAx	3,5275	0,0216	5,81%	0,61%
659	EgoAx	3,2859	0,0192	5,18%	0,59%
661	DirIndR , WheelAngle	2,0363	0,0184	4,94%	0,90%
664	DirIndL , WheelAngle	2,0390	0,0175	4,70%	0,86%
669	WheelAngle	1,8012	0,0157	4,24%	0,87%
675	DirIndR , EgoAy	2,0379	0,0145	3,91%	0,71%
676	DirIndL , WheelAngleDt	1,3972	0,0142	3,83%	1,02%
677	DirIndL , EgoAy	2,0429	0,0142	3,83%	0,70%

679	DirIndR , WheelAngleDt	1,3924	0,0140	3,78%	1,01%
683	EgoAy	1,7991	0,0129	3,47%	0,72%
684	DirIndR , GasPedal	4,5102	0,0127	3,43%	0,28%
685	DirIndL , GasPedal	4,5078	0,0126	3,41%	0,28%
687	WheelAngleDt	1,1542	0,0125	3,37%	1,09%
692	GasPedal	4,2671	0,0105	2,82%	0,25%
697	DirIndR , GasPedalDt	2,9359	0,0076	2,04%	0,26%
698	DirIndL , GasPedalDt	2,9361	0,0075	2,01%	0,25%
700	GasPedalDt	2,6896	0,0056	1,50%	0,21%
701	DirIndL , DirIndR	0,4596	0,0025	0,68%	0,55%
702	DirIndR	0,2483	0,0011	0,29%	0,43%
703	DirIndL	0,2491	0,0009	0,25%	0,38%

A.2 Maximum Relevance Ranking of Semantic Features

Table 10: Extract of the maximum relevance ranking of generated features from the video labeling sequence ontologies. Multiple entries with the same semantic – and thus mutual – information are in most cases omitted (e.g. TrafficSign, TrafficSign.isPart and TrafficSign.isPart.Road); some multiple entries are left for illustration (e.g. Vehicle.isRightOf and Vehicle.isRightOf.EgoCar).

Pos	Semantic Feature F	Entropy	1.8584	Label Road Type (Rd_...)	
		$H(F)$	$I(F;C)$	$I(F;C)/H(C)$	$I(F;C)/H(F)$
1	Lane	1.5794	0.6445	34.68%	40.80%
4	LaneLeft	0.9996	0.5773	31.07%	57.76%
7	isPart	2.2481	0.3563	19.17%	15.85%
10	EgoCar . isOn . LaneRight	0.9101	0.3046	16.39%	33.47%
11	LaneCenter	0.8155	0.2471	13.30%	30.30%
15	Vehicle . isRightOf	1.6155	0.2265	12.19%	14.02%
17	Vehicle . isRightOf . EgoCar	1.6155	0.2265	12.19%	14.02%
19	VehicleDualLane . isRightOf . EgoCar	1.5774	0.2245	12.08%	14.23%
21	VehicleDualLane . isRelPos . EgoCar	2.9673	0.1765	9.50%	5.95%
23	EgoCar . isOn . LaneCenter	0.6027	0.1660	8.93%	27.54%
24	VehicleDualLane	3.1913	0.1622	8.73%	5.08%
26	VehDual_Truck . isRelPos . EgoCar	1.3472	0.1612	8.68%	11.97%
28	VehDual_Truck . isRightOf . EgoCar	0.7608	0.1552	8.35%	20.40%
29	VehDual_Truck	1.3746	0.1544	8.31%	11.24%
33	Vehicle . isRelPos . EgoCar	3.0066	0.1507	8.11%	5.01%
35	VehicleDualBig . isRightOf . EgoCar	0.8857	0.1423	7.66%	16.06%
36	VehicleDualSmall	2.6775	0.1412	7.60%	5.27%
37	Vehicle	3.2421	0.1403	7.55%	4.33%
39	VehicleDualSmall . isRightOf . EgoCar	1.1833	0.1370	7.37%	11.58%
40	TrafficSign	1.6237	0.1278	6.88%	7.87%
44	VehicleDualSmall . isRelPos . EgoCar	2.5792	0.1214	6.53%	4.71%
46	EgoCar . isOn . LaneLeft	0.6703	0.1200	6.46%	17.90%
49	VehicleDualBig . isRelPos . EgoCar	1.5844	0.1162	6.25%	7.33%
51	VehDual_Van . isRelPos . EgoCar	0.9615	0.1096	5.90%	11.40%
52	VehDual_Car	2.4992	0.1060	5.70%	4.24%
54	VehDual_Van . isRightOf . EgoCar	0.4881	0.1019	5.48%	20.87%
55	VehicleDualBig	1.6182	0.1014	5.46%	6.27%
57	VehDual_Car . isRightOf . EgoCar	1.0186	0.0962	5.18%	9.44%
58	SpeedLimit	1.3967	0.0859	4.62%	6.15%
60	SpeedLimit . isPart . Road	1.3967	0.0859	4.62%	6.15%
61	VehicleSingleLane	0.9543	0.0846	4.56%	8.87%
62	SL70	0.5968	0.0846	4.55%	14.17%
66	VehDual_Car . isRelPos . EgoCar	2.4032	0.0839	4.52%	3.49%
68	VehicleDualLane . isLeftOf . EgoCar	1.9360	0.0799	4.30%	4.13%
72	Vehicle . isLeftOf . EgoCar	1.9624	0.0749	4.03%	3.82%

76	VehicleSingleLane . isInFrontOf . EgoCar	0.7094	0.0737	3.96%	10.38%
77	isOn	0.4875	0.0731	3.93%	14.99%
79	isOn . Lane	0.4875	0.0731	3.93%	14.99%
80	EgoCar . isOn . Lane	0.4875	0.0731	3.93%	14.99%
82	VehSgle_Motorbike . isInFrontOf . EgoCar	0.6952	0.0728	3.92%	10.47%
84	VehDual_Trailer . isInFrontOf . EgoCar	0.2208	0.0710	3.82%	32.15%
86	VehDual_Truck . isLeftOf . EgoCar	0.6330	0.0685	3.68%	10.81%
89	VehSgle_Motorbike . isRelPos . EgoCar	0.7924	0.0673	3.62%	8.50%
91	VehicleDualBig . isOncoming . EgoCar	0.8073	0.0654	3.52%	8.10%
93	VehicleDualSmall . isLeftOf . EgoCar	1.7127	0.0640	3.45%	3.74%
97	Vehicle . isInFrontOf . EgoCar	1.9757	0.0634	3.41%	3.21%
99	VehDual_Truck . isOncoming . EgoCar	0.6482	0.0625	3.37%	9.65%
101	VehicleDualLane . isOncoming . EgoCar	1.8060	0.0601	3.24%	3.33%
103	VehDual_Trailer . isRelPos . EgoCar	0.3596	0.0593	3.19%	16.49%
104	NoOvtakeAll	0.3795	0.0585	3.15%	15.42%
109	VehDual_Car . isLeftOf . EgoCar	1.6025	0.0566	3.04%	3.53%
110	SL100	0.4324	0.0562	3.03%	13.01%
114	VehDual_Car . isOncoming . EgoCar	1.4433	0.0561	3.02%	3.88%
118	Vehicle . isOncoming . EgoCar	1.8303	0.0556	2.99%	3.04%
120	VehicleSingleLane . isRelPos . EgoCar	0.8465	0.0506	2.72%	5.97%
122	VehicleDualSmall . isOncoming . EgoCar	1.5255	0.0499	2.68%	3.27%
123	VarSL	0.1936	0.0480	2.58%	24.78%
126	SL90	0.0941	0.0445	2.40%	47.33%
130	VehicleDualLane . isInFrontOf . EgoCar	1.8228	0.0445	2.39%	2.44%
131	NoOvertake	0.6404	0.0427	2.30%	6.67%
134	SL80	0.1833	0.0376	2.03%	20.54%
138	NoOvtakeAllEnd	0.1912	0.0343	1.84%	17.91%
141	SL30	0.1340	0.0320	1.72%	23.89%
144	StaticSL	1.1359	0.0320	1.72%	2.81%
148	VehDual_Bus . isOncoming . EgoCar	0.1480	0.0318	1.71%	21.50%
152	VehicleSingleLane . isRightOf . EgoCar	0.1768	0.0306	1.65%	17.31%
153	SL50	0.1951	0.0289	1.55%	14.79%
157	VehDual_Truck . isInFrontOf . EgoCar	0.5879	0.0286	1.54%	4.86%
159	VehDual_Van . isInFrontOf . EgoCar	0.4234	0.0281	1.51%	6.64%
160	StaticSLEnd	0.2977	0.0277	1.49%	9.29%
166	VehicleDualSmall . isInFrontOf . EgoCar	1.5944	0.0273	1.47%	1.71%
167	VarNoOvtakeLorry	0.1237	0.0269	1.45%	21.71%
170	VarSL100	0.1221	0.0264	1.42%	21.63%
173	ZoneSLEnd	0.0995	0.0252	1.36%	25.37%
174	ZoneSLEnd30	0.0995	0.0252	1.36%	25.37%
180	VehDual_Van . isLeftOf . EgoCar	0.4375	0.0246	1.32%	5.63%
181	SLCityEnd	0.1095	0.0244	1.32%	22.33%
184	NoOvtakeLorry	0.1219	0.0231	1.24%	18.96%
190	VehDual_Trailer . isRightOf . EgoCar	0.1415	0.0223	1.20%	15.75%
192	VehSgle_Motorbike . isRightOf . EgoCar	0.1084	0.0221	1.19%	20.40%
193	VehSgle_Bicycle	0.1539	0.0218	1.17%	14.14%
194	VarSL120	0.1015	0.0209	1.13%	20.63%
198	VehDual_Car . isInFrontOf . EgoCar	1.4642	0.0204	1.10%	1.39%
200	VehicleDualBig . isInFrontOf . EgoCar	0.8066	0.0203	1.09%	2.51%
201	SL120	0.1044	0.0190	1.02%	18.16%
205	VehDual_Trailer . isOncoming . EgoCar	0.2099	0.0189	1.02%	8.99%
208	SL60	0.1369	0.0163	0.88%	11.92%
217	VehSgle_Bicycle . isRelPos . EgoCar	0.0988	0.0142	0.76%	14.32%
219	VehSgle_Bicycle . isRightOf . EgoCar	0.0760	0.0138	0.74%	18.13%
220	SL110	0.0677	0.0138	0.74%	20.37%
223	AllLimitsEnd	0.0677	0.0135	0.73%	19.94%
228	VehDual_Bus . isRightOf . EgoCar	0.1278	0.0120	0.65%	9.42%
239	VehDual_Other	0.0765	0.0099	0.53%	12.89%
241	VehDual_Bus . isInFrontOf . EgoCar	0.0810	0.0096	0.51%	11.82%
243	VehDual_Van . isOncoming . EgoCar	0.3229	0.0095	0.51%	2.93%

246	VehDual_Other . isInFrontOf . EgoCar	0.0432	0.0091	0.49%	21.15%
247	VehDual_Other . isRelPos . EgoCar	0.0432	0.0091	0.49%	21.15%
248	SLEnd80	0.0496	0.0087	0.47%	17.60%
252	VehSgle_Motorbike . isOncoming . EgoCar	0.1249	0.0083	0.45%	6.62%
254	VehicleSingleLane . isOncoming . EgoCar	0.1309	0.0081	0.44%	6.20%
256	VehDual_Bus . isLeftOf . EgoCar	0.0822	0.0077	0.41%	9.33%
257	SLEnd70	0.0397	0.0071	0.38%	17.90%
260	SLEnd60	0.0582	0.0055	0.29%	9.38%
264	VehSgle_Motorbike . isLeftOf . EgoCar	0.1557	0.0046	0.25%	2.93%
266	VehicleSingleLane . isLeftOf . EgoCar	0.1612	0.0043	0.23%	2.68%
267	VehDual_Bus	0.2380	0.0030	0.16%	1.27%
269	VehDual_Bus . isRelPos . EgoCar	0.2343	0.0030	0.16%	1.29%
271	VehSgle_Bicycle . isInFrontOf . EgoCar	0.0189	0.0030	0.16%	15.73%
278	VehSgle_Bicycle . isLeftOf . EgoCar	0.0110	0.0019	0.10%	16.85%
279	VehSgle_Bicycle . isOncoming . EgoCar	0.0110	0.0019	0.10%	16.85%
283	EgoCar	0.0000	0.0000	0.00%	NaN
284	Road	0.0000	0.0000	0.00%	NaN
285	LaneRight	0.0000	0.0000	0.00%	NaN
288	EgoCar . isOn . Road	0.0000	0.0000	0.00%	NaN
289	LaneRight . isPart . Road	0.0000	0.0000	0.00%	NaN

References

[Althoff et al., 2009] Althoff, M., Stursberg, O., and Buss, M. (2009). Model-based probabilistic collision detection in autonomous driving. *IEEE Transactions on Intelligent Transportation Systems*, 10(2):299–310.

[Antoniou and van Harmelen, 2009] Antoniou, G. and van Harmelen, F. (2009). *Handbook on Ontologies*, chapter Web Ontology Language: OWL, pages 91–110. International Handbooks on Information Systems. Springer.

[Arens and Nagel, 2003] Arens, M. and Nagel, H.-H. (2003). Behavioral knowledge representation for the understanding and creation of video sequences. In *Proceedings of the 26th German Conference on Artificial Intelligence (KI-2003)*, pages 149–163.

[Baader et al., 2010] Baader, F., Calvanese, D., McGuinness, D. L., Nardi, D., and Patel-Schneider, P. F., editors (2010). *The Description Logic Handbook: Theory, Implementation and Applications*. Cambridge University Press, 2 edition.

[Baader and Nutt, 2009] Baader, F. and Nutt, W. (2009). *Handbook on Ontologies*, chapter Basic Description Logics, pages 47–104. International Handbooks on Information Systems. Springer.

[Bangso and Wuillemin, 2000] Bangso, O. and Wuillemin, P.-H. (2000). Object oriented bayesian networks: A framework for topdown specification of large bayesian networks and repetitive structures. Technical report, Aalborg University.

[Batina et al., 2010] Batina, L., Gierlichs, B., Prouff, E., Rivain, M., Standaert, F.-X., and Veyrat-Charvillon, N. (2010). Mutual Information Analysis: a Comprehensive Study. *Journal of Cryptology*, pages 1–23.

[Battiti, 1994] Battiti, R. (1994). Using mutual information for selecting features in supervised neural net learning. *Neural Networks, IEEE Transactions on*, 5(4):537–550.

[Baumgartner et al., 2008] Baumgartner, N., Retschitzegger, W., and Schwinger, W. (2008). Application scenarios of ontology-driven situation awareness systems - exemplified for the road traffic management domain. In *FOMI'08*, pages 77–87.

[BMVBS, 2009] BMVBS (2009). *Allgemeine Verwaltungsvorschrift zur Straßenverkehrs-Ordnung (VwV-StVO) vom 22. Oktober 1998 (General Directive about the Road Traffic Regulations)*. Bundesministerium für Verkehr, Bau und Stadtentwicklung (BMVBS), july 17, 2009 edition.

[Bonev et al., 2008] Bonev, B., Escolano, F., and Cazorla, M. (2008). Feature selection, mutual information, and the classification of high-dimensional patterns: Applications to image classification and microarray data analysis. *Pattern Analysis Applications*, 11(3-4):309–319.

[Brandt, 2004]Brandt, S. (2004). On subsumption and instance problem in elh w.r.t. general tboxes. In *International Workshop on Description Logics (DL2004)*, pages –1–1.

[Brechtel et al., 2011]Brechtel, S., Gindele, T., and Dillmann, R. (2011). Probabilistic mdp-behavior planning for cars. In *14th International IEEE Conference on Intelligent Transportation Systems (ITSC)*, pages 1537–1542.

[Breiman, 2001]   Breiman, L. (2001). Random forests. In *Machine Learning*, pages 5–32.

[Börger et al., 2010] Börger, J., Häring, J., and Palm, G. (2010). Verbesserte kursprädiktion durch maschinelles lernen am beispiel eines kollisionswarnsystems (improved course prediction through machine learning at the example of collision warning systems). In *Tagungsband zum 19. Aachener Kolloquium Fahrzeug- und Motorentechnik*.

[Broadhurst et al., 2005]   Broadhurst, A., Baker, S., and Kanade, T. (2005). Monte carlo road safety reasoning. In *IEEE Proceedings of Intelligent Vehicles Symposium*, pages 319–324.

[Brown, 2009]Brown, G. (2009). A new perspective for information theoretic feature selection. In *Twelfth International Conference on Artificial Intelligence and Statistics*.

[Bundesministerium der Justiz, Juris GmbH, 2009] Bundesministerium der Justiz, Juris GmbH, editor (2009). *Straßenverkehrs-Ordnung (StVO) (German traffic regulation law)*.

[Dasarathy, 1997]   Dasarathy, B. V. (1997). Sensor fusion potential exploitation – innovative architectures and illustrative applications. In *Proceedings of the IEEE*, volume 85, pages 24–38.

[Davis et al., 1993]   Davis, A., Shrobe, H., and Szolovits, P. (1993). What is a knowledge representation? *AI Magazine*, 14:17–33.

[DIN IEC 60050-351:2006, 2006]   DIN IEC 60050-351:2006 (2006). DIN IEC 60050-351:2006 international electrotechnical vocabulary - part 351: Control technology.

[Domingos et al., 2006]   Domingos, P., Kok, S., Poon, H., Richardson, M., and Singla, P. (2006). Unifying logical and statistical ai. In *Proceedings of the 21st National Conference on Artificial Intelligence (AAAI)*, pages 2–7. AAAI Press.

[Durak et al., 2006]   Durak, U., Oguztuzun, H., and Ider, S. K. (2006). An ontology for trajectory simulation. In *Proceedings of the 38th conference on Winter simulation*, WSC '06, pages 1160–1167. Winter Simulation Conference.

[Economic Commission for Europe: Inland Transport Committee, 1993] Economic Commission for Europe: Inland Transport Committee, editor (1993). *Convention on Road Traffic*.

[Endsley, 1995]   Endsley, M. R. (1995). Towards a theory of situation awareness in dynamic systems. *Human Factors*, 37:32.

[Endsley, 2000]   Endsley, M. R. (2000). Theoretical underpinnings of situation awareness: A critical review. In Endsley, M. R. and Garland, D. J., editors, *Situation
Awareness Analysis and measurement*. L. Erlbaum Assoc. Publishers.

[Estevez et al., 2009] Estevez, P., Tesmer, M., Perez, C., and Zurada, J. (2009). Normalized
mutual information feature selection. *Neural Networks, IEEE Transactions on*,
20(2):189 –201.

[European Commission, 2001]   European Commission, editor (2001). *White Paper - European Transport Policy for 2010: Time to Decide*. Office for Official Publica-
tions of the European Communities.

[European Commission, 2010]   European Commission (2010). Towards a european road
safety area: Policy orientations on road safety 2011-2020.

[European Commission – Information Society Technologies, 2008]   European Commission – Information Society Technologies (2008). Final report: Prevent – preventive and active safety applications integrated project (esafety for road and
air transport).

[FGSV, 2010] FGSV (2010). *Richtlinien für Lichtsignalanlagen (RiLSA): Lichtzeichenanla-
gen für den Straßenverkehr (Directives for Traffic Lights: Traffic Light Systems for Road Traffic)*. Forschungsgesellschaft für Straßen- und Verkehrswe-
sen, Arbeitsgruppe Verkehrsmanagement (FGSV).

[Frank and Asuncion, 2010] Frank, A. and Asuncion, A. (2010). UCI machine learning repository.

[Fraser and Swinney, 1986] Fraser, A. M. and Swinney, H. L. (1986). Independent coordi-
nates for strange attractors from mutual information. *Physical Review A*,
33(2):1134–1140.

[Freedman and Adams, 2007]   Freedman, S. T. and Adams, J. A. (2007). The inherent
components of unmanned vehicle situation awareness. In *Systems, Man and
Cybernetics, 2007. ISIC. IEEE International Conference on*, pages 973 –977.

[Fuchs, 2008] Fuchs, S. (2008). *A Comprehensive Knowledge Base for Context-Aware Tacti-
cal Driver Assistance Systems*. PhD thesis, Alpen-Adria-University Klagenfurt.

[Fuchs et al., 2008a] Fuchs, S., Rass, S., and Kyamakya, K. (2008a). Integration of ontological scene representation and logic-based reasoning for context-aware driver assistance systems. *ECEASST*.

[Fuchs et al., 2008b] Fuchs, S., Rass, S., Lamprecht, B., and Kyamakya, K. (2008b). A model for ontology-based scene description for context-aware driver assistance systems. *1st International ICST Conference on Ambient Media and Systems*.

[Giugno and Lukasiewicz, 2002]   Giugno, R. and Lukasiewicz, T. (2002). P-shoq(d): A
probabilistic extension of shoq(d) for probabilistic ontologies in the semantic
web. In *Proceedings of the European Conference on Logics in Artificial Intelligence*, JELIA '02, pages 86–97, London, UK, UK. Springer-Verlag.

[Gries et al., 2010]   Gries, O., Möller, R., Nafissi, A., Rosenfeld, M., Sokolski, K., and
Wessel, M. (2010). A probabilistic abduction engine for media interpretation
(extended version). Technical report, Hamburg University of Technology.

[Guyon et al., 2004] Guyon, I., Gunn, S. R., Ben-Hur, A., and Dror, G. (2004). Result analysis of the nips 2003 feature selection challenge. In *NIPS'04*, pages –1–1.

[Haarslev et al., 2011] Haarslev, V., Hidde, K., Möller, R., and Wessel, M. (2011). The racerpro knowledge representation and reasoning system. *SWJ Semantic Web Journal*, 2.

[Haarslev and Möller, 2001] Haarslev, V. and Möller, R. (2001). Racer system description. In Goré, R., Leitsch, A., and Nipkow, T., editors, *International Joint Conference on Automated Reasoning, IJCAR'2001, June 18-23, Siena, Italy*, pages 701–705. Springer-Verlag.

[Haarslev and Möller, 2004] Haarslev, V. and Möller, R., editors (2004). *International Workshop on Description Logics (DL2004)*. Whistler, BC, Canada.

[Häring and Wilhelm, 2009] Häring, J. and Wilhelm, U. (2009). Situation-interpretation as a key enabler for cost-effective and low-risk driver assistance systems with high collision mitigation capabilities. In *ESV 21st Enhanced Safety Vehicles Conference Proceedings*.

[Hermann and Desel, 2008] Hermann, A. and Desel, J. (2008). Driving situation analysis in automotive environment. In *IEEE International Conference on Vehicular Electronics and Safety (ICVES 2008)*, pages 216–221.

[Hesser, 2011]Hesser, J. (2011). Integration einer wissensbasis für fahrerassistenzsysteme am beispiel einer ampelassistenz (integration of a knowledge-base for driver assistance systems at the example of a traffic light assistance). Master's thesis, University of Stuttgart, Germany. German.

[Horrocks et al., 2003] Horrocks, I., Patel-Schneider, P. F., and van Harmelen, F. (2003). From shiq and rdf to owl: The making of a web ontology language. In *Web Semantics: Science, Services and Agents on the World Wide Web*, number 1, pages 7–26.

[Horrocks et al., 2000] Horrocks, I., Sattler, U., and Tobies, S. (2000). Reasoning with individuals for the description logic shiq. In *17th International Conferences on Automated Deduction (CADE-17)*, pages 482–496.

[Howard, 2010] Howard, C. (2010). *Knowledge Representation and Reasoning for a Model-Based Approach to Higher Level Information Fusion*. PhD thesis, University of South Australia. School of Computer and Information Science.

[Howard and Stumptner, 2005] Howard, C. and Stumptner, M. (2005). Situation assessments using object oriented probabilistic relational models. In *IEEE 8th International Conference on Information Fusion, 2005*, volume 2, page 8 pp.

[Howard and Stumptner, 2006] Howard, C. and Stumptner, M. (2006). Model construction algorithms for object-oriented probabilistic relational models. In *FLAIRS Conference*, pages 830–835.

[Howard and Stumptner, 2009] Howard, C. and Stumptner, M. (2009). Automated compilation of object-oriented probabilistic relational models. *International Journal on Approximate Reasoning*, 50:1369–1398.

[Hu et al., 2004]   Hu, W., Xiao, X., Xie, D., Tan, T., and Maybank, S. (2004). Traffic accident prediction using 3-d model-based vehicle tracking. *IEEE Transactions on Vehicular Technology 2004*, 53(3):677–694.

[Huang et al., 1994] Huang, T., Koller, D., Malik, J., Ogasawara, G., Rao, B., Russell, S., and Weber, J. (1994). Automatic symbolic traffic scene analysis using belief networks. In *AAAI'94: Proceedings of the 12th National Conference on Artificial Intelligence*, volume 2, pages 966–972, Menlo Park, CA, USA. American Association for Artificial Intelligence.

[Hülsen et al., 2010] Hülsen, M., Börger, J., and Zöllner, J. M. (2010). Ermittlung relevanter merkmale in komplexen verkehrssituationen (identification of relevant features in complex road traffic situations). In GmbH, V. W., editor, *26. VDI/VW-Gemeinschaftstagung Fahrerassistenz und Integrierte Sicherheit*, pages 123–138. VDI Verlag GmbH.

[Hülsen et al., 2011a] Hülsen, M., Zöllner, J. M., Häberlen, N., and Weiss, C. (2011a). Asynchronous real-time framework for knowledge-based intersection assistance. In *Intelligent Transportation Systems Conference (ITSC), 2011 IEEE*.

[Hülsen et al., 2011b] Hülsen, M., Zöllner, J. M., and Weiss, C. (2011b). Traffic intersection situation description ontology for advanced driver assistance. In *Intelligent Vehicles Symposium (IV), 2011 IEEE*, pages 993–999.

[Hummel, 2009]   Hummel, B. (2009). *Description Logic for Scene Understanding at the Example of Urban Road Intersections*. PhD thesis, Fakultät für Maschinenbau, Universität Karlsruhe (TH).

[Hummel et al., 2007] Hummel, B., Thiemann, W., and Lulcheva, I. (2007). Description logic for vision-based intersection understanding. In *Proc. Cognitive Systems with Interactive Sensors (COGIS), Stanford University, CA*.

[INTERSAFE-2, 2011]   INTERSAFE-2 (2011). Intersafe-2 – cooperative intersection safety (funded by the EUROPEAN COMMISSION).

[IPG Automotive, 2011]   IPG Automotive (2011). CarMaker – virtual test driving.

[Jaumard et al., 2006] Jaumard, B., Fortin, A., Shahriar, I., and Sultana, R. (2006). First order probabilistic logic. In *Annual meeting of the North American Fuzzy Information Processing Society (NAFIPS 2006)*, pages 341–346.

[Kammel et al., 2008] Kammel, S., Ziegler, J., Pitzer, B., Werling, M., Gindele, T., Jagzent, D., Schröder, J., Thuy, M., Goebl, M., Hundelshausen, F. v., Pink, O., Frese, C., and Stiller, C. (2008). Team annieway's autonomous system for the 2007 darpa urban challenge. *Journal of Field Robotics*, 25(9):615–639.

[Kane, 1989] Kane, T. B. (1989). Maximum entropy in nilsson's probabilistic logic. In *Proceedings of IJCAI 1989*. Morgan Kaufmann.

[Kate and Mooney, 2009]   Kate, R. J. and Mooney, R. J. (2009). Probabilistic abduction using markov logic networks. In *Proceedings of the IJCAI-09 Workshop on Plan, Activity, and Intent Recognition (PAIR)*.

[Keyarsalan and Montazer, 2010]   Keyarsalan, M. and Montazer, G. (2010). Intelligent on-
    tological agent for traffic light control of isolated intersections. In *Digital Con-
    tent, Multimedia Technology and its Applications (IDC), 2010 6th Internation-
    al Conference on*, pages 34 –39.

[Knaup and Homeier, 2010] Knaup, J. and Homeier, K. (2010). Roadgraph - graph based en-
    vironmental modelling and function independent situation analysis for driver
    assistance systems. In *Intelligent Transportation Systems (ITSC), 2010 13th In-
    ternational IEEE Conference on*, pages 428 –432.

[Kokar et al., 2009]   Kokar, M. M., Matheus, C. J., and Baclawski, K. (2009). Ontology-
    based situation awareness. *Information Fusion*, 10(1):83 – 98. Special Issue on
    High-level Information Fusion and Situation Awareness.

[Kwak and Choi, 1999]   Kwak, N. and Choi, C. (1999). Improved mutual information
    feature selector for neural networks in supervised learning. In *Neural Net-
    works, 1999. IJCNN '99. International Joint Conference on*, volume 2, pages
    1313 –1318 vol.2.

[Kwak and Choi, 2002a]   Kwak, N. and Choi, C. (2002a). Input feature selection by mutu-
    al information based on parzen window. *Pattern Analysis and Machine Intelli-
    gence, IEEE Transactions on*, 24(12):1667–1671.

[Kwak and Choi, 2002b]   Kwak, N. and Choi, C.-H. (2002b). Input feature selection for
    classification problems. *Neural Networks, IEEE Transactions on*, 13(1):143 –
    159.

[Lattner et al., 2005] Lattner, A. D., Gehrke, J. D., Timm, I. J., and Herzog, O. (2005). A
    knowledge-based approach to behavior decision in intelligent vehicles. In *IEEE
    Proceedings of the Intelligent Vehicles Symposium 2005*, pages 466–471.

[Liu and Hu, 2009]   Liu, C.-T. and Hu, B.-G. (2009). Mutual information based on renyi's
    entropy feature selection. In *Intelligent Computing and Intelligent Systems,
    2009. ICIS 2009. IEEE International Conference on*, volume 1, pages 816 –
    820.

[Liu et al., 2008a]   Liu, F., Sparbert, J., and Stiller, C. (2008a). Immpda vehicle tracking
    system using asynchronous sensor fusion of radar and vision. In *IEEE Intelli-
    gent Vehicles Symposium 2008*, pages 168–173.

[Liu et al., 2008b]   Liu, H., Liu, L., and Zhang, H. (2008b). Feature selection using mutual
    information: An experimental study. In *PRICAI '08: Proceedings of the 10th
    Pacific Rim International Conference on Artificial Intelligence*, pages 235–
    246, Berlin, Heidelberg. Springer-Verlag.

[Llinas et al., 2004] Llinas, J., Bowman, C., Rogova, G., Steinberg, A., Waltz, E., and Whi-
    te, F. (2004). Revisiting the jdl data fusion model ii. In Svensson, I. P. and
    Schubert, J., editors, *Proceedings of the 7th International Conference on In-
    formation Fusion (FUSION 2004)*, pages 1218–1230.

[Lloyd, 1987] Lloyd, J. W. (1987). *Foundations of Logic Programming (Symbolic Computa-
    tion / Artificial Intelligence)*. Springer, 2nd edition.

[Meyer-Delius et al., 2008] Meyer-Delius, D., Plagemann, C., Wichert, G., Feiten, W., Lawitzky, G., and Burgard, W. (2008). A probabilistic relational model for characterizing situations in dynamic multi-agent systems. In *Data Analysis, Machine Learning and Applications*, pages 269–276.

[Méndez et al., 2006] Méndez, J. R., Fdez-Riverola, F., Díaz, F., Iglesias, E. L., and Corchado, J. M. (2006). A comparative performance study of feature selection methods for the anti-spam filtering domain. In *Industrial Conference on Data Mining'06*, pages 106–120.

[Montemerlo et al., 2008] Montemerlo, M., Becker, J., Bhat, S., Dahlkamp, H., Dolgov, D., Ettinger, S., Haehnel, D., Hilden, T., Hoffmann, G., Huhnke, B., Johnston, D., Klumpp, S., Langer, D., Levandowski, A., Levinson, J., Marcil, J., Orenstein, D., Paefgen, J., Penny, I., Petrovskaya, A., Pflueger, M., Stanek, G., Stavens, D., Vogt, A., and Thrun, S. (2008). Junior: The stanford entry in the urban challenge. *J. Field Robot.*, 25:569–597.

[Naeth and Möller, 2008] Naeth, T. and Möller, R. (2008). Implementing probabilistic description logics: An application to image interpretation. In *Dagstuhl Seminar Proceedings: Logic and Probability for Scene Interpretation*.

[Nardi and Brachman, 2009] Nardi, D. and Brachman, R. J. (2009). *Handbook on Ontologies*, chapter An Introduction to Description Logics, pages 1–44. International Handbooks on Information Systems. Springer.

[NAVTEQ NN4D, 2011] NAVTEQ NN4D (2011). NAVTEQ Network for Developers (NN4D) – ADAS Research Platform (ADAS RP).

[NHTSA (National Highway Traffic Safety Administration), 2011] NHTSA (National Highway Traffic Safety Administration) (2011). Vehicle safety rulemaking and research priority plan for 2011–2013.

[Nienhüser et al., 2011] Nienhüser, D., Gumpp, T., and Zöllner, J. M. (2011). Relevance estimation of traffic elements using markov logic networks. In *14th International IEEE Conference on Intelligent Transportation Systems (ITSC)*, pages 1659–1664.

[Nilsson, 1986] Nilsson, N. J. (1986). Probabilistic logic. In *Artificial Intelligence*, volume 28, pages 71–87. Elsevier Science.

[Pellkofer, 2003] Pellkofer, M. (2003). *Verhaltensentscheidung für autonome Fahrzeuge mit Blickrichtungssteuerung*. PhD thesis, Universität der Bundeswehr München.

[Pellkofer and Dickmanns, 2002] Pellkofer, M. and Dickmanns, E. D. (2002). Behavior decision in autonomous vehicles. In *IEEE Intelligent Vehicle Symposium 2002*, volume 2, pages 495–500 vol.2.

[Peng et al., 2005] Peng, H., Long, F., and Ding, C. (2005). Feature selection based on mutual information: Criteria of max-dependency, max-relevance, and min-redundancy. *Pattern Analysis and Machine Intelligence, IEEE Transactions on*, 27(8):1226–1238.

[Polychronopoulos et al., 2006]   Polychronopoulos, A., Amditis, A., Scheunert, U., and
     Tatschke, T. (2006). Revisiting jdl model for automotive safety applications:
     The pf2 functional model. In *9th International Conference on Information Fu-sion. ICIF*, pages 1–7.

[Pommerening et al., 2009]   Pommerening, F., Wölfl, S., and Westphal, M. (2009). Right-of-way rules as use case for integrating golog and qualitative reasoning. In
     Mertsching, B., Hund, M., and Aziz, Z., editors, *KI 2009: Advances in Artifi-cial Intelligence*, volume 5803 of *Lecture Notes in Computer Science*, pages
     468–475. Springer Berlin / Heidelberg.

[Regele, 2008]   Regele, R. (2008). Using ontology-based traffic models for more effi-cient decision making of autonomous vehicles. In *Autonomic and Autonomous
     Systems, 2008. ICAS 2008. Fourth International Conference on*, pages 94 –99.

[Reif, 2010]   Reif, K. (2010). *Fahrstabilisierungssysteme und Fahrerassistenzsysteme (Dri-ving Stabilisation Systems and Driver Assistance Systems)*, volume 1.
     Vieweg+Teubner, Wiesbaden.

[Richardson and Domingos, 2006]   Richardson, M. and Domingos, P. (2006). Markov logic
     networks. In *Machine Learning*, page 2006.

[Robert Bosch GmbH, 2011a]   Robert Bosch GmbH, editor (2011a). *Automotive Hand-book*. Wiley, 8 edition.

[Robert Bosch GmbH, 2011b]   Robert Bosch GmbH, editor (2011b). *Kraftfahrtechni-sches Taschenbuch (German Edition)*. Vieweg+Teubner Verlag, 27 edition.

[Roy, 2001]   Roy, J. (2001). From data fusion to situation analysis. In *Proceedings of Fu-sion 2001 International Conference, Montréal, Canada*.

[Russell and Norvig, 2003]   Russell, S. and Norvig, P. (2003). *Artificial Intelligence: A
     Modern Approach*. Prentice Hall, second edition.

[Salerno, 2002]   Salerno, J. (2002). Information fusion: A high-level architecture over-view. In *Proceedings of the Fifth International Conference on Information Fu-sion 2002*, volume 1, pages 680–686 vol.1.

[Schamm and Zöllner, 2011] Schamm, T. and Zöllner, J. M. (2011). A model-based approach
     to probabilistic situation assessment for driver assistance systems. In *14th In-ternational IEEE Conference on Intelligent Transportation Systems (ITSC)*,
     pages 1404–1409.

[Schmidt et al., 2009] Schmidt, G. J., Khanafer, A., and Balzer, D. (2009). Successive catego-rization of perceived urgency in dynamic driving situations. In *Proceedings of
     the SAE World Congress & Exhibition*.

[Schneider et al., 2008]   Schneider, J., Wilde, A., and Naab, K. (2008). Probabilistic ap-proach for modeling and identifying driving situations. In *IEEE Intelligent Ve-hicles Symposium 2008*, pages 343–348.

[Schuricht et al., 2011]   Schuricht, P., Michler, O., and Bäker., B. (2011). Efficiency-increasing driver assistance at signalized intersections using predictive traffic
     state estimation. In *14th International IEEE Conference on Intelligent Trans-portation Systems (ITSC)*.

[Skutek et al., 2005] Skutek, M., Linzmeier, D. T., Appenrodt, N., and Wanielik, G. (2005). A precrash system based on sensor data fusion of laser scanner and short range radars. In *8th International Conference on Information Fusion 2005*, volume 2, pages 8 pp.–.

[Sowa, 1987] Sowa, J. F. (1987). Semantic networks. In *Encyclopedia of Artificial Intelligence*. Wiley.

[Sowa, 1992] Sowa, J. F. (1992). Semantic networks (revised and extended). In *Encyclopedia of Artificial Intelligence*. Wiley, 2 edition.

[Sowa, 1999] Sowa, J. F. (1999). *Knowledge Representation: Logical, Philosophical, and Computational Foundations*. Brooks / Cole, 1 edition.

[Spinda, 2011] Spinda, J. (2011). Wissensbasierte situationsinterpretation für fahrerassistenzsysteme (knowledge-based situation interpretation for driver assistance systems). Master's thesis, University of Albstadt-Sigmaringen, Germany. German.

[Srinivasa, 2005] Srinivasa, S. (2005). A review on multivariate mutual information. *EE-80653 Information Theory Tutorials*.

[Steinberg et al., 1999] Steinberg, A. N., Bowman, C. L., and White, F. E. (1999). Revisions to the jdl data fusion model. In Dasarathy, B. V., editor, *Sensor Fusion: Architectures, Algorithms, and Applications III*, volume 3719, pages 430–441. SPIE.

[Stiller et al., 2007] Stiller, C., Farber, G., and Kammel, S. (2007). Cooperative cognitive automobiles. In *IEEE Intelligent Vehicles Symposium 2007*, pages 215–220.

[Torkkola et al., 2004] Torkkola, K., Venkatesan, S., and Liu, H. (2004). Sensor selection for maneuver classification. In *Intelligent Transportation Systems, 2004. Proceedings. The 7th International IEEE Conference on*, pages 636 – 641.

[Tourassi et al., 2001] Tourassi, G. D., Frederick, E. D., Markey, M. K., and Floyd, C. E. J. (2001). Application of the mutual information criterion for feature selection in computer-aided diagnosis. *Med Phys.*, 28:2394–2402.

[Urmson et al., 2008] Urmson, C., Anhalt, J., Bae, H., Bagnell, J. D., Baker, C., Bittner, R. E., Brown, T., Clark, M. N., Darms, M., Demitrish, D., Dolan, J., Duggins, D., Ferguson, D., Galatali, T., Geyer, C. M., Gittleman, M., Harbaugh, S., Hebert, M., Howard, T., Kolski, S., Likhachev, M., Litkouhi, B., Kelly, A., McNaughton, M., Miller, N., Nickolaou, J., Peterson, K., Pilnick, B., Rajkumar, R., Rybski, P., Sadekar, V., Salesky, B., Seo, Y.-W., Singh, S., Snider, J. M., Struble, J. C., Stentz, A. T., Taylor, M., Whittaker, W. R. L., Wolkowicki, Z., Zhang, W., and Ziglar, J. (2008). Autonomous driving in urban environments: Boss and the urban challenge. In *Journal of Field Robotics (JFR) Special Issue on the 2007 DARPA Urban Challenge, Part I*, volume 25, pages 425–466.

[Vacek et al., 2007] Vacek, S., Gindele, T., Zöllner, J. M., and Dillmann, R. (2007). Using case-based reasoning for autonomous vehicle guidance. In *IEEE/RSJ International Conference on Intelligent Robots and Systems (IROS 2007)*, pages 4271–4276.

[van Harmelen et al., 2008]   van Harmelen, F., Lifschitz, V., and Porter, B., editors (2008). *Handbook of Knowledge Representation (Foundations of Artificial Intelligence)*. Elsevier Science.

[VIRES, 2011]   VIRES (2011). Opendrive (vires simulationstechnologie gmbh).

[Weiser, 2010]   Weiser, A. (2010). A probabilistic lane change prediction module for highly automated driving. In *Proceedings of the 7th International Workshop on Intelligent Transportation (WIT 2010)*.

[Yang and Moody, 1999]   Yang, H. H. and Moody, J. (1999). Feature selection based on joint mutual information. In *In Proceedings of International ICSC Symposium on Advances in Intelligent Data Analysis*, pages 22–25.

[Zhang et al., 2006]   Zhang, X., Zhao, H., Sun, P., and Xu, Y. (2006). An improved algorithm for reducing bayesian network inference complexity. In *Signal Processing, 2006 8th International Conference on*, volume 4.